SET THEORY AND TOPOLOGY

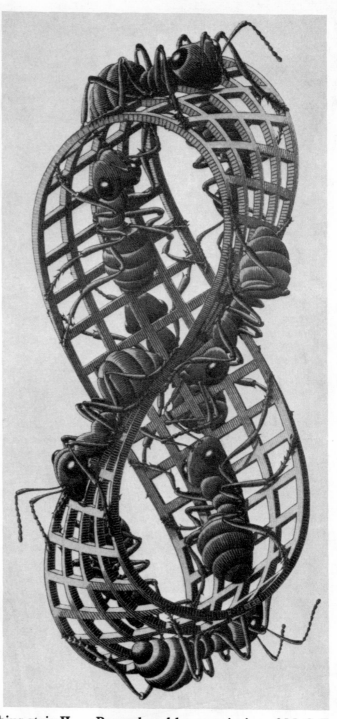

Moebius strip II   Reproduced by permission of M. C. Escher

SET THEORY
AND TOPOLOGY

Philip Nanzetta
St. Mary's College of Maryland

George E. Strecker
University of Florida

Bogden & Quigley, Inc.
Publishers

Tarrytown-on-Hudson, New York / Belmont, California

1971

To the Thylacine, *Thylacinus cynocephalus*,
who is almost extinct

QA
248
N28

Cover designed by Winston G. Potter

Text designed by Science Bookcrafters, Inc.

Library of Congress Catalog Card No.: 71–132070

Standard Book No.: –0–8005–0004–0

Printed in the United States of America

1 2 3 4 5 6 7 8 9 10 — 75 74 73 72 71

PREFACE

It is common for a graduate or advanced undergraduate mathematics text to specify "mathematical maturity" as the major prerequisite for its use. This book does not have that requirement. Indeed, it is designed for use in a course whose chief goal is precisely the *development* of mathematical maturity. This goal is achieved by increasing student participation, specifically by having students provide their own proofs of theorems.

It has been our experience that a one-year course conducted in this fashion is extremely valuable. Such a course permits students to learn what a proof is and how proofs are discovered. This early experience of constructing proofs not only helps to sharpen the students' critical faculties, but it also allows participation in the excitement of discovering proofs and gives an appreciation for the mathematics that will be confronted later. Practice in presenting one's own work in front of peers also serves as a confidence builder and improves the ability to communicate.

Because this text is designed for use in a course that does not employ the lecture-style format, a great deal of care and attention has been given to motivation of the topics presented. To a large extent, we believe that it is self-motivating and in most instances there is a smooth flow from each topic to the next. This, together with the fact that set theory and topology lend themselves particularly well to the do-it-yourself approach, has allowed us to cover nearly as much material using this text as is covered in similar courses that use the traditional lecture method.

Chapter 1 introduces notions from the algebra of sets, including cartesian products, equivalence relations, and cardinality. There are several reasons for initially approaching these concepts from an intuitive standpoint. The axiomatic theory that appears later is much easier to understand after first becoming familiar with the ideas to be abstracted. The need for a more rigorous approach can then be driven home by means of a contradiction. Introducing the intuitive theory first also allows one to develop manipulative skill in a less exacting setting, so that by the time one considers the axioms, the proofs of a great many of the early theorems are familiar, and concentration can then be focused on details involved with the axioms. Finally, this approach allows a quite rapid treatment of the fundamental set-theoretic concepts so that this groundwork does not have to be duplicated in the beginning stages of other courses which may be taken concurrently.

Chapter 2 deals first with the axioms for Gödel–Bernays–von Neumann set theory and with propositions that do not require the axiom of choice. A sequence of statements equivalent to the axiom of choice is presented, followed by some applications. The remainder of the chapter is devoted to a development of ordinal and cardinal numbers. Enough of the arithmetic of ordinal and cardinal numbers is covered so that a student will gain the familiarity with it that he will need in his future everyday mathematics. Appendices B and C are given to provide a brief but rather complete tabular summary of the entire set-theoretic development.

Based only on the axioms for set theory given in Chapter 2, Chapter 3 provides a brief construction of the real numbers and develops some of their essential properties.

Chapter 4 presents general topology in its abstract setting. However, it is motivated throughout by familiar examples from calculus and the geometry of two- and three-dimensional euclidean spaces. Also, full use is made of the students' recently acquired background in set theory. The chapter presents the standard basic material expected in a first graduate course in topology, including continuity, compactness, connectedness, separation axioms, product and quotient spaces, metrizable spaces, completeness, the Tychonoff product theorem, the Stone–Čech compactification, the Baire category theorems, Urysohn's lemma, Tietze's extension theorem, and Urysohn's metrization theorem.

Even though we have included no category theory in the text, seeds of category-theoretical ideas (such as duality and universality) have been planted throughout. In this way a student will more easily grasp the general concepts when he eventually confronts them.

We wish to express our appreciation to Professors Kermit Sigmon and David Stadtlander, who, together with the authors, wrote class notes that were forerunners of those on which this book is based. Gratitude is also due our students, whose suggestions and criticisms proved to be invaluable in the development of this material. We owe a further debt to Professors Don Plank, Charles Saalfrank, and Neil Wagner, who read the completed manuscript and offered numerous suggestions for its improvement. As usual we claim for ourselves all deficiencies in our presentation.

We were most fortunate to be able to locate a competent typist. Without this stroke of fortune, the writing of this book would have been much more painful.

P.N.
G.E.S.

TO THE INSTRUCTOR

"Teaching" from this book for the first time is likely to be a memorable experience. You forget just how difficult it is to sit quietly and watch someone present a proof different from the one you have in mind, or one that is too sketchy or is burdensomely detailed. But this pain is ultimately worth it. The reward of actually seeing a student who didn't even know what a proof was at the beginning present a beautiful, polished proof after some months of work justifies the pain on your part and the effort on his. Patience is called for, and criticism, and trick questions, and traps. Blind alleys must be followed to the end. But most of the time you must sit and be quiet.

We feel that your task is that of a moderator and, to some extent, an encyclopedia. If more illustrative examples are needed at a particular spot, you should provide them. Occasionally, further hints will be needed, but you should give them only if no one in the class can help. You should point out ways to improve proofs and ways to improve presentations. You should control the amount of detail expected in proofs and examples (and, in Chapter 2, which of the repeated theorems should be reproved). But mostly you should moderate. Don't tell a student when he turns into a blind alley; let him find out for himself. Only correct an error if the rest of the class has let it go—and encourage them not to let errors go. In what we conceive to be the ideal situation, you should hardly say a word in class, except to expand on something. This ideal can only be approximated, but the approximation improves as the course progresses.

CONTENTS

CHAPTER 1 INTUITIVE SET THEORY

We begin the study of set theory by examining its intuitive content. The need for a more formal approach, exemplified by a system of axioms, will become evident later.

A **set** or **class** is any collection of elements. For example, the collection of all university students, of all red shoes, or of all integers less than 10 or greater than 80 will be sets. Sets may be specified by any meaningful formula or property. If P is any such property and $P(x)$ is the statement that x has property P, then $\{x \mid P(x)\}$ denotes the class of all x such that $P(x)$ is true. For example, if $Q(y)$ is the statement "y is red and y is a shoe", then $\{y \mid Q(y)\}$ is the set of all red shoes. The third example above can be written $\{x \mid x$ is an integer, and $x < 10$ or $x > 80\}$. If a set has only finitely many members, say $a_1, a_2, \ldots, a_n$, then it may be written $\{a_1, a_2, \ldots, a_n\}$; this is really just shorthand for $\{x \mid x = a_1$ or $\cdots$ or $x = a_n\}$.* Membership will be denoted by the symbol $\in$. Thus "$x \in A$" means "x belongs to A" or "x is an element of A" or "x is a member of A"; "$x \notin A$" means "x is not a member of A".

It is often convenient to consider the set with no members. This set is called the **empty set** or **null set** and is denoted by $\varnothing$. Note that there is nothing wrong with sets themselves being elements of other sets (e.g., the set of all lines in the plane where each line is the set of its points). If A is a set, then $\{A\}$ (read **singleton** A) denotes the set whose sole member is A. Thus if $\mathbf{Z}$ is the set of all integers, then $\mathbf{Z}$ has infinitely many elements, whereas $\{\mathbf{Z}\}$ has only one; $\varnothing$ has no members, but $\{\varnothing\}$ has one and $\{\varnothing, \{\varnothing\}\}$ has two. Two sets A and B are said to be **equal** (denoted $A = B$) provided that they have precisely the same elements. Thus $A = B$ means that "A" and "B" are just two names for the same collection of elements, or more explicitly it means "for all x, $x \in A$ if and only if $x \in B$". $A \neq B$ means that A and B are not equal.

1.1 Definition

Let A be a set.

(a) B will be called a **subset** of A (written $B \subseteq A$, "B is **contained in** A") iff† every member of B is a member of A; i.e., $B \subseteq A$ means that for all x, $x \in B$ implies $x \in A$. $B \nsubseteq A$ means that B is not contained in A. $B \supseteq A$ means $A \subseteq B$.

(b) B will be called a **proper subset** of A (denoted by $B \subsetneqq A$) iff $B \subseteq A$ and $B \neq A$.

---

* Throughout this book (and indeed all of mathematics), the statement "a or b" means "a or b or both".

† "**iff**" is an abbreviation for the phrase "if and only if".

1.2 **Examples**

(a) For every set A, $A \subseteq A$.

(b) For any sets A and B, $A = B$ iff $A \subseteq B$ and $B \subseteq A$.

(c) Let **N** denote the set of positive integers; **E**, the set of even integers; **Z**, the set of integers; **Q**, the set of rational numbers; and **R**, the set of real numbers.   Then $\mathbf{N} \subsetneqq \mathbf{Z} \subsetneqq \mathbf{Q} \subsetneqq \mathbf{R}$, but $\mathbf{N} \nsubseteq \mathbf{E}$ and $\mathbf{E} \nsubseteq \mathbf{N}$.

(d) $\varnothing$ is a subset of every set and a proper subset of every set except $\varnothing$ itself.

Definition 1.1 provides a method of constructing sets: If A is a set, we can consider the set of all subsets of A, denoted by $\mathscr{P}(A)$ (and called the **power set** of A).   Thus $\mathscr{P}(A) = \{B \mid B \subseteq A\}$.   Note that for every set A, we have $A \in \mathscr{P}(A)$ and $\varnothing \in \mathscr{P}(A)$.

1.3 **Example**

$\mathscr{P}(\{\varnothing, \{\varnothing\}\}) = \{\varnothing, \{\varnothing\}, \{\{\varnothing\}\}, \{\varnothing, \{\varnothing\}\}\}$.

1.4 **Definition**

If A and B are sets, then the **complement of B in A**, denoted by $A - B$, is the set of all elements of A that do not belong to B; i.e., $A - B = \{x \mid x \in A \text{ and } x \notin B\}$.

1.5 **Examples**

(a) For every set A, $A - A = \varnothing$ and $A - \varnothing = A$.

(b) $\mathscr{P}(A) - \{A\}$ is the set of all proper subsets of A.

(c) $\mathbf{N} - \mathbf{E}$ is the set of all odd positive integers.

1.6 **Proposition**

If A is a subset of a set C, then $C - (C - A) = A$.

PROOF $C - (C - A) = \{x \mid x \in C \text{ and } x \notin C - A\}$

$\qquad\qquad = \{x \mid x \in C \text{ and it is not true that } (x \in C \text{ and } x \notin A)\}$

$\qquad\qquad = \{x \mid x \in C \text{ and it is not true that } x \notin A\}$

$\qquad\qquad = \{x \mid x \in C \text{ and } x \in A\}$

$\qquad\qquad = A,$

the last equality holding since $A \subseteq C$.

1.7 **Proposition**

If A and B are subsets of a set C, then $A \subseteq B$ iff $C - A \supseteq C - B$.

PROOF   Suppose $A \subseteq B$.   Let $x \in C - B$.   Since $x \in B$ if $x \in A$, we must have $x \notin A$. Since $x \in C$, we have $x \in C - A$.   Thus $A \subseteq B$ implies $C - A \supseteq C - B$.

Suppose $C - A \supseteq C - B$. Then by what has been shown and by 1.6, $A = C - (C - A) \subseteq C - (C - B) = B$.

1.8   Definition

If $\mathscr{S}$ is a set of sets, then the **union** of $\mathscr{S}$, denoted by $\bigcup \mathscr{S}$, is the set $\{x \mid \text{for some } A \in \mathscr{S},\ x \in A\}$. If $\mathscr{S} = \{A, B\}$, then $\bigcup \mathscr{S}$ is usually written $A \cup B$ and read "A **union** B".

1.9   Examples

(a) For any set A, $A \cup A = A \cup \varnothing = A$.

(b) For any sets A, B, and C, $A \cup B = \{x \mid x \in A \text{ or } x \in B\}$, $A \cup B = B \cup A$, and $(A \cup B) \cup C = A \cup (B \cup C) = \bigcup \{A, B, C\}$.

(c) For any sets A and B, $A \subseteq B$ iff $A \cup B = B$.

(d) $\bigcup \varnothing = \varnothing$.

1.10   Proposition

(a) If $\mathscr{S} = \{\{x\} \mid x \in B\}$, then $\bigcup \mathscr{S} = B$.

(b) If $\mathscr{S} = \{\{B\}\}$, then $\bigcup(\bigcup \mathscr{S}) = B$.

PROOF   (a) Let $x \in B$. Then $x \in \{x\} \in \mathscr{S}$, so by the definition of union, $x \in \bigcup \mathscr{S}$. Hence $B \subseteq \bigcup \mathscr{S}$. Let $x \in \bigcup \mathscr{S}$. Then there is some $A \in \mathscr{S}$ such that $x \in A$. Since every element of $\mathscr{S}$ is a set with exactly one element, $A = \{x\}$. Hence, since $\{x\} \in \mathscr{S}$, we have $x \in B$. Thus $\bigcup \mathscr{S} \subseteq B$. These inclusions, together with Example 1.2(b), complete the proof.

(b) From now on, you should supply all missing proofs. Selected hints are given beginning on page 101.

1.11   Definition

If $\mathscr{S}$ is a nonempty set of sets, then the **intersection** of $\mathscr{S}$, denoted by $\bigcap \mathscr{S}$, is the set $\{x \mid \text{for all } A \in \mathscr{S},\ x \in A\}$. If $\mathscr{S} = \{A, B\}$, then $\bigcap \mathscr{S}$ is usually written $A \cap B$ and is read "A **intersect** B". If $A \cap B = \varnothing$, then A and B are said to be **disjoint**; if $A \cap B \neq \varnothing$, then we say that A **meets** B.

1.12   Examples

(a) For any set A, $A \cap A = A$ and $A \cap \varnothing = \varnothing$.

(b) For any sets A, B, and C, $A \cap B = \{x \mid x \in A \text{ and } x \in B\}$; $A \cap B = B \cap A$; $(A \cap B) \cap C = A \cap (B \cap C) = \bigcap \{A, B, C\}$.

(c) For any sets A and B, $A \subseteq B$ iff $A \cap B = A$.

(d) For any sets A and B, $A = A \cap (A \cup B) = A \cup (A \cap B)$.

(e) For any sets A and B, $B - (B - A) = B \cap A$.

1.13

The proper use of "pictures" can often greatly simplify the process of discovering a set-theoretic fact. We wish to stress that pictures do not constitute a proof, but rather are tools which make its discovery easier.

We illustrate the use of **Venn diagrams**. If A is represented by the inside of one blob and B by the inside of another, then the intersection of A and B is the part common to the two blobs, as shown in Figure 1.1. The

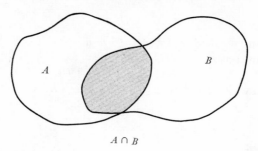

$A \cap B$

Figure 1.1

union of A and B is the part that is inside either blob, as shown in Figure 1.2. The complement of B in A is the part inside the A blob and outside the B blob, as shown in Figure 1.3.

By comparing Figures 1.1 and 1.3 it is easy to see, for example, that $A - B = A - (A \cap B)$. We suggest that you always construct (at least mentally) the Venn diagram associated with any proof that you attempt.

Of course it may happen that a Venn diagram is so poorly drawn that it suggests a false result. For example, Figure 1.4 might lead one to believe that for all sets A, B, and C, $A \cap B \cap (C - A) = A \cap B \cap C$. This is why an actual proof must be constructed even after a picture suggesting the proof has been drawn.

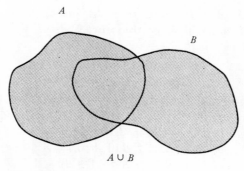

$A \cup B$

Figure 1.2

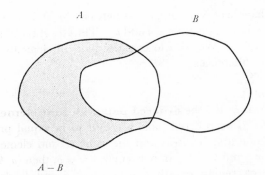

$A - B$

Figure 1.3

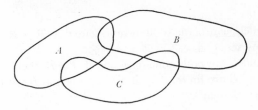

Figure 1.4

1.14   Proposition
Let A, B, and C be sets.   Then
   (a) $A - (B \cup C) = (A - B) \cap (A - C)$.
   (b) $A - (B \cap C) = (A - B) \cup (A - C)$.

This proposition is called **De Morgan's laws**, and it has the following more general form.

1.15   Proposition
Let A be a set and $\mathscr{S}$ be a nonempty set of sets.   Then
   (a) $A - (\bigcup \mathscr{S}) = \bigcap \{A - C \,|\, C \in \mathscr{S}\}$.
   (b) $A - (\bigcap \mathscr{S}) = \bigcup \{A - C \,|\, C \in \mathscr{S}\}$.

1.16   Proposition
Let A be a set and $\mathscr{S}$ be a set of sets.   Then
   (a) $A \cap (\bigcup \mathscr{S}) = \bigcup \{A \cap C \,|\, C \in \mathscr{S}\}$.
   (b) $A \cup (\bigcap \mathscr{S}) = \bigcap \{A \cup C \,|\, C \in \mathscr{S}\}$, if $\mathscr{S}$ is not empty.

1.17   Corollary
For any sets A, B, and C,
   (a) $A \cap (B \cup C) = (A \cap B) \cup (A \cap C)$.
   (b) $A \cup (B \cap C) = (A \cup B) \cap (A \cup C)$.

Recall that there is no difference between $\{a, b\}$ and $\{b, a\}$ since either is just the set whose elements are a and b. Thus $\{a, b\}$ is called the **unordered pair** of a and b. We wish to consider pairs of elements now where order *does* make a difference.

1.18   Definition

(a) (a, b) is called the **ordered pair** with **first element** a and **second element** b. Two ordered pairs are said to be equal provided that they have the same first elements and the same second elements; i.e., $(a, b) = (c, d)$ iff $a = c$ and $b = d$. In particular, if $a \neq b$, then $(a, b) \neq (b, a)$.

(b) The **cartesian product** of two sets A and B, denoted by $A \times B$, is the set $\{(a, b) \mid a \in A \text{ and } b \in B\}$.

1.19   Examples

(a) The euclidean plane is often represented as $\mathbf{R} \times \mathbf{R}$.

(b) For any set A, $A \times \varnothing = \varnothing \times A = \varnothing$.

(c) If A and B are both nonempty and if $A \neq B$, then $A \times B \neq B \times A$.

(d) If A and B are finite, A with m elements and B with n elements, then $A \times B$ has $m \cdot n$ elements.

1.20   Proposition

For any sets A, B, and C, $A \times (B - C) = (A \times B) - (A \times C)$.

1.21   Proposition

For any sets $\mathscr{S}$ and $\mathscr{T}$ of sets,

(a) $\bigcup \mathscr{S} \times \bigcup \mathscr{T} = \bigcup \{A \times B \mid (A, B) \in \mathscr{S} \times \mathscr{T}\}$.

(b) $\bigcap \mathscr{S} \times \bigcap \mathscr{T} = \bigcap \{A \times B \mid (A, B) \in \mathscr{S} \times \mathscr{T}\}$, if $\mathscr{S}$ and $\mathscr{T}$ are nonempty.

1.22   Corollary

For any sets A, B, C, and D,

(a) $A \times (B \cup C) = (A \times B) \cup (A \times C)$.

(b) $A \times (B \cap C) = (A \times B) \cap (A \times C)$.

1.23   Definition

Let A and B be sets. A **relation from A to B** is a subset of $A \times B$. A relation from A to A is called a **relation on A**. If R is a relation from A to B, then

$$\pi_1[R] = \{a \in A \mid \text{for some } b \in B, (a, b) \in R\}$$

is called the **domain** of R and

$$\pi_2[R] = \{b \in B \mid \text{for some } a \in A, (a, b) \in R\}$$

is called the **range** of R.   Note that if R is a relation from A to B, then $R \subseteq \pi_1[R] \times \pi_2[R]$. If $(a, b) \in R$ (sometimes denoted by $a\,R\,b$), then we say that a is R-**related** to b.   The relation $R^{-1} = \{(b, a) \mid (a, b) \in R\}$ is called the **inverse** of R.

1.24   Examples
(a) For any sets A and B, $\varnothing$ is a relation from A to B.

(b) If A is any set, then the **identity** relation or **diagonal** relation on A is $1_A = \Delta_A = \{(a, a) \mid a \in A\}$; A is the domain of as well as the range of 1_A; $\Delta_A = \Delta_A{}^{-1}$.

(c) The usual $<$ and $\leq$ on $\mathbf{R}$ can be thought of as relations (sketch them). What are $(<)^{-1}$ and $(\leq)^{-1}$?

(d) $S = \{(r, r^2) \mid r \in \mathbf{R}\}$ is a relation on $\mathbf{R}$.   Note that $S \neq S^{-1}$.

1.25   Definition
If R is a relation from A to B and if $C \subseteq A$, then

(a) the **restriction of** R **to** C is the relation from C to B defined by
$$R \mid C = \{(c, b) \mid (c, b) \in R \text{ and } c \in C\}$$
$$= R \cap (C \times B).$$

(b) the **image of** C **under** R is the set
$$R[C] = \{b \mid (c, b) \in R \text{ for some } c \in C\}$$
$$= \pi_2[R \mid C].$$

1.26   Definition
If R and S are relations, then the **composition** of R with S (or R **followed by** S) is the relation

$$S \circ R = \{(a, b) \mid \text{there exists a } c \text{ such that } (a, c) \in R \text{ and } (c, b) \in S\}.$$

1.27   Examples
(a) If R and S are relations, and if $\pi_2[R]$ and $\pi_1[S]$ are disjoint, then $S \circ R = \varnothing$.

(b) If $R \subseteq A \times B$, then $\Delta_B \circ R = R = R \circ \Delta_A$.

1.28   Proposition
Let R, S, and T be relations and let $\mathscr{S}$ be a set of relations.   Then

(a) $P \subseteq R$ and $Q \subseteq S$ implies that $P \circ Q \subseteq R \circ S$.

(b) $(R \circ S) \circ T = R \circ (S \circ T)$.

(c) $((R)^{-1})^{-1} = R$.

(d) $(R \circ S)^{-1} = S^{-1} \circ R^{-1}$.

(e) $R \circ (\bigcup \mathscr{S}) = \bigcup \{R \circ U \mid U \in \mathscr{S}\}$.

(f) $R \circ (\bigcap \mathscr{S}) \subseteq \bigcap \{R \circ U \mid U \in \mathscr{S}\}$, if $\mathscr{S}$ is nonempty.

(g) The containment in (f) cannot be replaced by equality.

1.29   Definition

A relation F from A to B is called a **function** provided that for every $a \in A$ there is one and only one $b \in B$ such that $(a, b) \in F$. This unique b is usually denoted by $F(a)$. Thus $F = \{(a, F(a)) \mid a \in A\}$. The notation $F : A \to B$ or $A \xrightarrow{F} B$ means that F is a function from A to B. Sometimes a function is defined by merely telling (by means of the tailed arrow "$\mapsto$") what it does to each of the elements of its domain; e.g., $r \mapsto r^2$ defines the function S of Example 1.24.

A function $F : A \to B$ is **onto** B (or is **surjective**) iff $F[A] = B$; i.e., for every $b \in B$, there is some $a \in A$ such that $b = F(a)$. The function $F : A \to B$ is **one-to-one** (or **injective**) provided that $F(a) \neq F(b)$ whenever $a \neq b$. A function $F : A \to B$ that is one-to-one and onto B is called a **bijection**.

1.30   Examples

(a) For every set X, $\Delta_X = 1_X$ is a bijection from X to X.

(b) For sets A and B, the empty relation $\varnothing$ from A to B is a function iff $A = \varnothing$. In this case it is called the **empty function**.

(c) If $F : A \to \varnothing$, then $A = \varnothing$ and $F = 1_\varnothing = \varnothing$.

(d) If $A \subseteq B$, then $i : A \to B$ defined by $i = 1_B \mid A$ is a one-to-one function called the **inclusion function of A into B**.

(e) In Example 1.24, S is a function from $\mathbf{R}$ to $\mathbf{R}$ that is not one-to-one and not onto.

(f) $\leq$, $<$, and S^{-1} of Example 1.24 are not functions.

(g) Any relation R from A to B induces a function $\hat{R} : \mathscr{P}(A) \to \mathscr{P}(B)$ defined by $\hat{R} = \{(C, R[C]) \mid C \in \mathscr{P}(A)\}$.

(h) For every cartesian product $A \times B$, there exist functions

$$\pi_1 : A \times B \to A \text{ defined by } (a, b) \mapsto a,$$

$$\pi_2 : A \times B \to B \text{ defined by } (a, b) \mapsto b.$$

π_1 (resp., π_2) is called the **first** (resp., **second**) **projection function** of $A \times B$.* Both are surjections iff $A \times B \neq \varnothing$ or $A = B = \varnothing$.

(i) Every cartesian product also admits a bijection from $A \times B$ to $B \times A$ defined by $(a, b) \mapsto (b, a)$.

1.31   Proposition

If f and g are functions, then $f \circ g$ is a function and for every x in the domain of $f \circ g$, $(f \circ g)(x) = f(g(x))$.

---

* Note that if $R \subseteq A \times B$, there is no ambiguity in the meaning of $\pi_1[R]$ considered as the domain of the relation R (1.23), and considered as the image of the set R under the first projection function. Similarly for $\pi_2[R]$.

1.32   **Proposition**
Let $F : A \to B$.   The following statements are equivalent.
(a) F is one-to-one.
(b) $F^{-1} \circ F = 1_A$.
(c) If $A \neq \emptyset$, then there exists some $G : B \to A$ such that $G \circ F = 1_A$.
(d) For every set C and all pairs of functions $H : C \to A$ and $K : C \to A$ such that $F \circ H = F \circ K$, it follows that $H = K$ (i.e., F is **left-cancellable** with respect to composition).
(e) F^{-1} is a function from $F[A]$ to A.

PROOF   We will show that (a) $\Rightarrow$ (b) $\Rightarrow$ (c) $\Rightarrow$ (d) $\Rightarrow$ (e) $\Rightarrow$ (a).
(a) $\Rightarrow$ (b)

$$F^{-1} \circ F = \{(a, c) \mid \text{there exists a } b \text{ such that } (a, b) \in F \text{ and } (b, c) \in F^{-1}\}$$
$$= \{(a, c) \mid \text{there exists a } b \text{ such that } (a, b) \in F \text{ and } (c, b) \in F\}$$
$$= \{(a, c) \mid a = c\}$$
$$= 1_A;$$

the third equality holds since F is assumed to be a one-to-one function.
(b) $\Rightarrow$ (c)   Let $a_0 \in A$.   Let G be the relation

$$G = F^{-1} \cup \{(b, a_0) \mid b \in B - F[A]\}.$$

Let (b, a) and (b, c) belong to G. If $b \in F[A]$, then $(a, b) \in F$ so that $(a, c) \in F^{-1} \circ F = 1_A$; hence $a = c$. If $b \in B - F[A]$, then $a = a_0 = c$. Thus, in any event $a = c$, so that G is a function.
Since $F^{-1} \circ F = 1_A$ and the domain of $\{(b, a_0) \mid b \in B - F[A]\}$ is disjoint from the range of F, we have $G \circ F = 1_A$.
(c) $\Rightarrow$ (d)   If $A = \emptyset$, then $H = \emptyset = K$.   Suppose that $A \neq \emptyset$ and G is the function given by part (c).   If $F \circ H = F \circ K$, then $G \circ (F \circ H) = G \circ (F \circ K)$,   so that $(G \circ F) \circ H = (G \circ F) \circ K$.   Since $G \circ F = 1_A$,   this yields $H = 1_A \circ H = 1_A \circ K = K$.
(d) $\Rightarrow$ (e)   Suppose (a, b) and (a, c) belong to F^{-1}. Let $C = \{x\}$ and let $H = \{(x, b)\}$ and $K = \{(x, c)\}$.   Then $F \circ H = F \circ K = \{(x, a)\}$.   Hence $H = K$, so that $b = c$.   Thus F^{-1} is a function.
(e) $\Rightarrow$ (a)   Suppose $F(a) = F(b)$.   Then $(F(a), a) \in F^{-1}$ and $(F(a), b) = (F(b), b) \in F^{-1}$.   Since F^{-1} is a function, $a = b$.

1.33   **Proposition**
Let $F : A \to B$.   The following statements are equivalent.
(a) F is onto B.
(b) $F \circ F^{-1} = 1_B$.
(c) There exists some $G : B \to A$ such that $F \circ G = 1_B$.

(d) For every set C and all pairs of functions $H : B \to C$ and $K : B \to C$ such that $H \circ F = K \circ F$ it follows that $H = K$ (i.e., F is **right-cancellable** with respect to composition).

1.34   Corollary

A function $F : A \to B$ is a bijection iff there exists $G : B \to A$ such that $F \circ G = 1_B$ and $G \circ F = 1_A$.

1.35   Proposition

Let R be a relation from A to B, $\mathscr{S} \subseteq \mathscr{P}(A)$, and $\varnothing \neq \mathscr{T} \subseteq \mathscr{P}(B)$.   Then
(a) $R[\bigcup \mathscr{S}] = \bigcup \{R[C] \mid C \in \mathscr{S}\}$.
(b) $R[\bigcap \mathscr{S}] \subseteq \bigcap \{R[C] \mid C \in \mathscr{S}\}$, if $\mathscr{S} \neq \varnothing$.
(c) If R is a function, then $R^{-1}[\bigcap \mathscr{T}] = \bigcap \{R^{-1}[D] \mid D \in \mathscr{T}\}$.
(d) The containment in (b) cannot be replaced by equality.

1.36

The notion of the cartesian product of two sets is easy to generalize (by means of n-tuples) to the case in which there is a finite number of factors.   For example, $A \times B \times C$ can be taken to be the set of all ordered triples of the form (a, b, c), where $a \in A$, $b \in B$, and $c \in C$.   However, problems arise when there are infinitely many "factor" sets.   A satisfactory generalization is given below (1.39).

1.37   Proposition

Let A_1 and A_2 be sets.   Let

$$P = \{F \mid F : \{1, 2\} \to A_1 \cup A_2, \ F(1) \in A_1 \text{ and } F(2) \in A_2\}.$$

Let $\hat{\pi}_1 : P \to A_1$ be defined by $\hat{\pi}_1(F) = F(1)$, and let $\hat{\pi}_2 : P \to A_2$ be defined by $\hat{\pi}_2(F) = F(2)$. Then there exists a unique bijection $H : P \to A_1 \times A_2$ such that the diagram

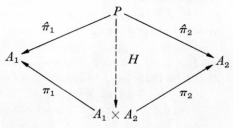

commutes, i.e., such that $\hat{\pi}_1 = \pi_1 \circ H$ and $\hat{\pi}_2 = \pi_2 \circ H$.

Proposition 1.37 shows that there is no essential set-theoretic difference between P together with the functions $\hat{\pi}_1$ and $\hat{\pi}_2$, and $A_1 \times A_2$ together with the functions π_1 and π_2.   This gives a clue to the proper generalization of the concept of cartesian product.

_1.38_ **Definition**

A **family of sets indexed by** a set I is a function F with domain I and range some set of sets. Usually for each $i \in I$, $F(i)$ is written F_i and the family is denoted by $(F_i)_{i \in I}$, $(F_i)_I$, or (F_i), depending upon how much notation is needed in any particular case. The image of I under F is usually written $\{F_i \mid i \in I\}$.

Notice that a family need not be a one-to-one function, so that it can happen that $A_i = A_j$ when $i \neq j$. Indeed, the only reason for defining a family to be a function rather than just a set of sets is that with a family a given set can be repeated and each repetition will "count".

_1.39_ **Definition**

Let $(A_i)_I$ be a family of sets indexed by I. Then the **cartesian product** of $(A_i)_I$ is the set

$$\Pi(A_i)_{i \in I} = \{F \mid F : I \to \bigcup\{A_i \mid i \in I\} \text{ such that for every } i \in I, F(i) \in A_i\}.$$

This is sometimes denoted more briefly by ΠA_i. Under these circumstances it is clear that if each A_i is nonempty, then for each $j \in I$, $\{(F, F(j)) \mid F \in \Pi A_i\}$ is a function from ΠA_i onto A_j (prove it); this function is denoted by $\pi_j : \Pi A_i \to A_j$ and is called the jth **projection** of ΠA_i or the **projection of ΠA_i onto its jth coordinate set.**

_1.40_ **Examples**

(a) Let A_1 and A_2 be sets. Then $\Pi(A_i)_{\{1,2\}}$ is the set P of Proposition 1.37 and is thus essentially the same as $A_1 \times A_2$.

(b) Let $(A_i)_I$ be a family of sets. If $I = \varnothing$, then $\Pi(A_i)_I = \{\varnothing\}$. If $I \neq \varnothing$ and one of the sets A_i is empty, then $\Pi(A_i)_I = \varnothing$. If each A_i is a singleton, then $\Pi(A_i)_I$ is a singleton.

_1.41_ **Theorem**

Let $(A_i)_I$ be a family of sets.

(a) The cartesian product ΠA_i together with projections $\pi_i : \Pi A_i \to A_i$ is "universal" in the following sense: For every set Y together with functions $f_i : Y \to A_i$ for each $i \in I$, there exists a unique function $F : Y \to \Pi A_i$ such that for each $i \in I$, the diagram

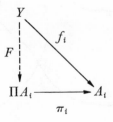

commutes.

(b) The above universality characterizes the cartesian product in the following sense: If Z is a set together with functions $\rho_i : Z \to A_i$ for each $i \in I$ such that for every set Y together with functions $f_i : Y \to A_i$ for each $i \in I$, there exists a unique function $F : Y \to Z$ such that for each $i \in I$, the diagram

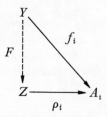

commutes; then Z is essentially the cartesian product ΠA_i; i.e., there exists a bijection $H : Z \to \Pi A_i$ such that for each $i \in I$, the diagram

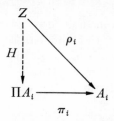

commutes.

1.42

It is often the case that given sets A_1 and A_2, we wish to form their "disjoint union" or "sum", i.e., the union of A_1 and A_2, where A_1 and A_2 are considered to be disjoint.

For example, if we have sets A_1 and A_2 in the Venn diagram of Figure 1.5, their union $A_1 \cup A_2$ would be as shown in Figure 1.6, whereas we want their disjoint union $A_1 \uplus A_2$ to be as shown in Figure 1.7.

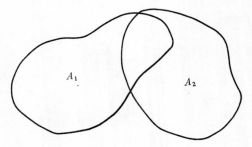

Figure 1.5

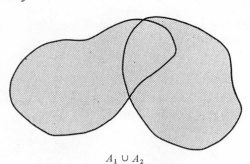

$$A_1 \cup A_2$$

Figure 1.6

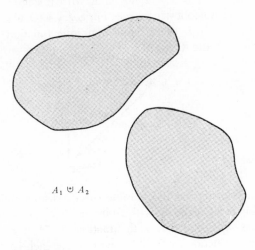

$$A_1 \uplus A_2$$

Figure 1.7

To accomplish this technically requires a method of making disjoint "copies" of A_1 and A_2. This can be done by attaching to each set a distinguishing factor (like painting one set red and the other blue) before forming the union; for example, we can define

$$A_1 \uplus A_2 = (A_1 \times \{1\}) \cup (A_2 \times \{2\}).$$

Then $A_1 \times \{1\}$ is "just like" A_1 (i.e., there exists a bijection $f\colon A_1 \to A_1 \times \{1\}$), $A_2 \times \{2\}$ is "just like" A_2, and $(A_1 \times \{1\})$ and $(A_2 \times \{2\})$ are disjoint.

1.43   **Definition**
Let $(A_i)_I$ be a family of sets indexed by I. Then the **disjoint union** of $(A_i)_I$ is the set

$$\sum (A_i)_I = \bigcup \{A_i \times \{i\} \mid i \in I\},$$

sometimes denoted more briefly by $\sum A_i$. Under these circumstances it is clear that for each $j \in I$, $a \mapsto (a, j)$ defines a one-to-one function from A_j to

$\sum A_i$; it is denoted by $\mu_j : A_j \to \sum A_i$ and is called the **injection of the** j**th coordinate set into** $\sum A_i$. If $I = \{1, 2, \ldots, n\}$, then $\sum (A_i)_I$ is often denoted by $A_1 \uplus A_2 \uplus \cdots \uplus A_n$.

1.44   Example

There is a set P such that for any set A there is a family $(P_i)_I$ with $P_i = P$ for all i and a bijection from A to $\sum (P_i)_I$. There is no analogous result for products.

1.45   Theorem

Let $(A_i)_I$ be a family of sets.

(a) The disjoint union $\sum A_i$ together with the injections $\mu_i : A_i \to \sum A_i$ is "universal" in the following sense: For every set Y together with functions $f_i : A_i \to Y$ for all $i \in I$, there exists a unique function $F : \sum A_i \to Y$ such that for each $i \in I$, the diagram

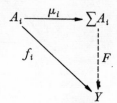

commutes.

(b) The above universality characterizes the disjoint union in the following sense: If Z is a set together with functions $\delta_i : A_i \to Z$ for each $i \in I$ such that for every set Y together with functions $f_i : A_i \to Y$ for each $i \in I$, there exists a unique function $F : Z \to Y$ such that for each $i \in I$, the diagram

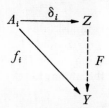

commutes; then Z is essentially the disjoint union $\sum A_i$; i.e., there exists a bijection $H : \sum A_i \to Z$ such that for each $i \in I$ the diagram

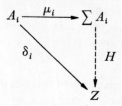

commutes.

1.46   **Definition**

Let R be a relation on a set A.

(a) R is **reflexive on** A iff $R \supseteq \Delta_A$, i.e., iff $(a, a) \in R$ for every $a \in A$.

(b) R is **symmetric** iff $R = R^{-1}$, i.e., iff $(b, a) \in R$ whenever $(a, b) \in R$.

(c) R is **transitive** iff $R \circ R \subseteq R$, i.e., iff $(a, b) \in R$ and $(b, c) \in R$ together imply that $(a, c) \in R$.

(d) R is a **quasi-order relation on** A iff R is reflexive on A and transitive.

(e) R is an **equivalence relation on** A iff R is a symmetric, transitive, and reflexive relation on A.

1.47   **Examples**

(a)   For any set A, the quasi-order relation

$$\subseteq \; = \{(B, C) \mid B \subseteq C \text{ and } B, C \in \mathscr{P}(A)\}$$

on $\mathscr{P}(A)$ is not an equivalence relation on $\mathscr{P}(A)$ if $A \neq \varnothing$.

(b)   The usual $\leq$ on $\mathbf{R}$ is a quasi-order relation on $\mathbf{R}$, but it is not symmetric.

(c)   For any set A, Δ_A is an equivalence relation on A.

(d)   For any function $F : A \to B$, $\{(a_1, a_2) \mid F(a_1) = F(a_2)\}$ is an equivalence relation on A called the **kernel** of F.

(e)   The relation R on $\mathbf{Z} \times \mathbf{N}$ defined by $(a, b) \; R \; (c, d)$ iff $ad = cb$ is an equivalence relation.

(f)   Let $n \in \mathbf{N}$.   Then the relation S on $\mathbf{N}$ defined by $(p, q) \in S$ iff p is congruent to q modulo n is an equivalence relation.

1.48   **Definition**

(a)   If R is an equivalence relation on A, and $a \in A$, then $R[\{a\}]$ is called the **equivalence class of** a **relative to** R.   The set $\{R[\{a\}] \mid a \in A\}$ of all equivalence classes relative to R is usually denoted by A/R (and is called "A **modulo** R" or "A mod R").

(b)   A **partition** of a set A is a set $\mathscr{D}$ of nonempty disjoint sets with the property that $\bigcup \mathscr{D} = A$.

1.49   **Theorem**

Let A be a set. If R is an equivalence relation on A, then A/R is a partition of A.   Conversely, if $\mathscr{D}$ is a partition of A, then $\bigcup \{B \times B \mid B \in \mathscr{D}\}$ is an equivalence relation Q on A and $A/Q = \mathscr{D}$.

Thus there is no essential difference between the partitions of a set and the equivalence relations on that set.

1.50   **Proposition**

If R is an equivalence relation on A, then $\eta_R = \{(a, R[\{a\}]) \mid a \in A\}$ is a surjective function from A to A/R.

1.51   Definition
(a) The function $\eta_R : A \to A/R$ is called the **quotient** (or **natural** or **canonical**) **mapping** of A onto A/R.
(b) If S is a subset of A for which $\eta_R \mid S$ is one-to-one and onto A/R, then S is called a **system of representatives** for the equivalence relation R.

1.52   Proposition
Every equivalence relation on a set has a system of representatives.

1.53   Definition
Sets A and B are said to be **equipotent** (denoted by $A \sim B$) iff there exists a bijection from A to B.

1.54   Proposition
$\sim$ is an equivalence relation.

1.55   Definition
For each set A, *card* A denotes the equivalence class under $\sim$ to which A belongs; i.e., *card* $A = \sim [\{A\}]$.

1.56   Proposition
Let A, A', B, and B' be sets with $A \sim A'$ and $B \sim B'$. Then A is equipotent with a subset of B iff A' is equipotent with a subset of B'.

This allows the following definition.

1.57   Definition
Let A and B be sets. Then *card* $A \leq$ *card* B iff A is equipotent with a subset of B.

1.58   Theorem
For any set A, *card* $\mathscr{P}(A) \nleq$ *card* A.

1.59   Proposition
$\leq$ is a quasi-order relation on $\{card\ A \mid A \text{ is a set}\}$.

1.60   Theorem
Let $\mathfrak{U} = \{A \mid A \text{ is a set}\}$. For every set A, *card* $A \leq$ *card* $\mathfrak{U}$.

1.61   Corollary
Intuitive set theory is inconsistent.

1.62

The problem here is buried in the proof of Theorem 1.58. If $\mathfrak{U} = \{A \mid A \text{ is a set}\}$ is permitted to be a set, then 1.58 is false, since $\mathscr{P}(\mathfrak{U}) \subseteq \mathfrak{U}$. On further analysis, the difficulty reduces to a well-known paradox of intuitive set theory—Russell's paradox—which goes as follows: Let $S = \{x \mid x \notin x\}$. Then if S is a set (as it would be in intuitive set theory), $S \in S$ implies that $S \notin S$ and $S \notin S$ implies that $S \in S$.

The problem here and in all standard paradoxes of this type is that sets are permitted to be too large. The way out is indicated by the parenthetical statement in the above paragraph: don't let S be a set (and don't let non-sets be members of sets). To accomplish this, we divide our intuitive "sets" into two categories: sets and "sets" that are too large; the latter things we call proper classes. More accurately, the "sets" of our intuitive theory will from now on be called classes and the well-behaved classes will be the sets of the axiomatic theory. We will still be able to manipulate classes to some extent, but proper classes (i.e., classes that are not sets) will not be allowed to be members of other classes. This approach avoids the more obvious paradoxes of intuitive set theory and is felt to be free of contradiction. (Cultural note: It was proved in 1931 by Gödel that *no system of axioms for an adequate set theory can be proved to be consistent.*) It is reasonable to ask why we must turn to axioms: Why not just say what has been said and proceed on the (more pleasant) intuitive level. The answer is that rather fine distinctions must be made at certain points and it is necessary to have the full contract in writing to get comfortably over these points. From an esthetic point of view, it *is* nice to have all assumptions made explicit at the beginning, and not to have to wonder exactly what can be assumed.

It is pleasant to know that all our efforts so far have not been wasted. After we present a system of axioms, we will find that *when we restrict our attention to sets*, all the theorems presented in sections before 1.54 will remain true with essentially the same proofs.

CHAPTER **2** AXIOMATIC SET THEORY

We take the ideas of **class** and **membership** ($\in$) as undefined notions (whose "meaning" comes entirely from the system of axioms below, but which are based on and motivated by the intuitive theory of sets); "$A \in B$" is read "A is a **member** of (or is an **element** of) B". By convention, *script letters will denote classes and roman letters will be used only for those classes which are sets*. Equality is taken to mean identity, so that "$\mathscr{A} = \mathscr{B}$" means that "$\mathscr{A}$" and "$\mathscr{B}$" are just different names for the same class.

Our axioms fall into three categories: statements of the existence of classes of certain types, statements that classes of certain types are sets, and statements restricting the nature of classes.

2.1 Definition

A class $\mathscr{A}$ is called a **set** iff there exists a class $\mathscr{B}$ with $\mathscr{A} \in \mathscr{B}$. A class that is not a set is called a **proper class**.

2.2 Axiom 1

If $\mathscr{A}$ and $\mathscr{B}$ are classes, then $\mathscr{A} = \mathscr{B}$ is equivalent to the statement "for all sets X, $X \in \mathscr{A}$ iff $X \in \mathscr{B}$".

Axiom 1 is known as the **axiom of extensionality** (origin: the "extension" of a class is the "collection" of its elements; the axiom says that two classes are equal iff they have the same extension).

2.3 Definition

If $\mathscr{A}$ and $\mathscr{B}$ are classes, we write $\mathscr{A} \subseteq \mathscr{B}$ iff for all sets X, $X \in \mathscr{A}$ implies $X \in \mathscr{B}$. If $\mathscr{A} \subseteq \mathscr{B}$, we say that $\mathscr{A}$ is a **subclass** of $\mathscr{B}$ or that $\mathscr{A}$ is **contained in** $\mathscr{B}$. If $\mathscr{A}$ is also a set, we say that $\mathscr{A}$ is a **subset** of $\mathscr{B}$. $\mathscr{A} \subsetneqq \mathscr{B}$ means $\mathscr{A} \subseteq \mathscr{B}$ and $\mathscr{A} \neq \mathscr{B}$.

2.4

In intuitive set theory we let any property specify a set by characterizing its elements. Unfortunately, this cannot be preserved: $\{X \mid X \notin X \text{ and } X \text{ is a set}\}$ is not a set, by Russell's paradox. On the other hand, we will be able to permit any reasonable property to specify a *class*. The purpose of the next few paragraphs is to make precise what is meant by "reasonable property".

A **primitive propositional function** (**ppf**) is any meaningful formula built up out of set and class variables and constants, the membership symbol, the logical operations "implies", "or", "and", and "not", and the so-called quantifiers "for all" and "there exists"—where the quantifiers are to apply *only to set variables*. Specifically:

(a) If $\mathscr{A}$ and $\mathscr{B}$ denote class or set variables or constant classes, then "$\mathscr{A} \in \mathscr{B}$" is a ppf.

(b) If P and Q are ppfs, then so are "P implies Q", "P or Q", "P and Q", and "not Q".

(c) If P is a ppf, then "there exists a set X such that P" and "for all sets X, P" are ppfs.

(d) If P is a ppf, then the formula obtained by replacing every occurrence of X in P by another set variable is a ppf.

(e) These are the only ppfs.

2.5 *Examples*

(a) The following statements are ppfs:
   (1) $X \notin X$ [we write "$X \notin X$" for "not $(X \in X)$"].
   (2) $(X \in \mathscr{A})$ or $((y \in X)$ and for all sets $Z, y \notin Z)$.
   (3) $(\mathscr{A} \in \mathscr{B}$ implies $Z \in Y)$ and there exists a set W such that (for all sets $y, y \in W$ implies $y \in \mathscr{B})$.

(b) The following strings of symbols are not ppfs:
   (1) For all classes $\mathscr{A}, y \notin \mathscr{A}$.
   (2) $(x \in Y_1)$ or $\cdots$ or $(x \in Y_n)$ or $\cdots$.
   (3) $\mathscr{A} \in$ not $\mathscr{B}$.

As we become accustomed to ppfs, we shall frequently fail to write out a statement in such a way that it is immediately obvious that it is a ppf; when in doubt, however, you should check to make sure. For example [after we have defined $\cup$ and $\mathscr{P}(\ \ )$] we will write such things as "$(X \in \mathscr{A} \cup \mathscr{B})$ and $X \subseteq \mathscr{P}(C)$" and use them as ppfs. The translation of this statement is "$((X \in \mathscr{A})$ or $(X \in \mathscr{B}))$ and (for all sets Y, $Y \in X$ implies (for all sets Z, $Z \in Y$ implies $Z \in C))$". It is clear why we do not reduce statements to their elementary form unless it is absolutely necessary.*

2.6

We need to place one further restriction on our formulas to attain reasonableness. It is hard to imagine how the expression $\{x \mid x \in A$ or $y \in B\}$ is to make sense if both x and y are variables. On the other hand, $\{x \mid x \in A$ or there exists y such that $y \in B\}$ seems to make sense. The distinction here is that x is the only variable in the second ppf that is free of the quantifiers while both x and y are free in the first.

Definition

(a) If X is a variable, then X is free in "$X \in \mathscr{A}$" and in "$\mathscr{A} \in X$".

(b) If X is a variable and X is free in one of the ppfs P or Q, then X is free

---

* Appendix B consists of a rigorous treatment of ppfs in tabular form. It can be used to aid in the reduction of standard set-theoretic statements to ppfs.

in "*P* or *Q*", "*P* and *Q*", and "*P* implies *Q*"; if *X* is free in *P*, then *X* is free in "not *P*".

(c) If *X* and *Y* are distinct set variables and if *X* is free in the ppf *P*, then *X* is free in "for all *Y*, *P*" and in "there exists a *Y* such that *P*".

(d) If *X* is a variable and *P* is a ppf, then *X* is **free in** *P* iff it can be proved free using (a), (b), and (c).   Otherwise, *X* is said to be **bound in** *P*.

2.7   *Examples*

(a) *X* is free in each of the following ppfs:
  (1) $(X \notin A)$ or $(X \in \mathscr{B})$.
  (2) $X \subseteq \mathscr{A}$.
  (3) $(X \in A)$ or (for all X, $X \in B$).
  (4) For all y, $(X = X$ and $y \in Z)$.

(b) *X* is bound in the following ppfs:
  (1) For all X, $X \subseteq \mathscr{A}$.
  (2) There exists an X such that $X \in Y$.

2.8   *Axiom 2*

If $P(X)$ is a ppf whose only free variable is X, then there exists a class whose members are precisely those *sets* X for which $P(X)$ holds.

Axiom 2 is called the **axiom of class formation**. We will usually denote by $\{X \mid P(X)\}$ the class whose existence is guaranteed by this axiom.

2.9   *Proposition*

If $\mathscr{A}$ is any class, then there exists a class [denoted by $\mathscr{P}(\mathscr{A})$] whose elements are exactly the subsets of $\mathscr{A}$.

PROOF   By 2.4(a) "$z \in X$" and "$z \in \mathscr{A}$" are ppfs, and by 2.6(a) *X* is free in "$z \in X$". Thus by 2.4(b) and 2.6(b) "$z \in X$ implies $z \in \mathscr{A}$" is a ppf in which *X* is free, so that by 2.4(c) and 2.6(c) "for all sets z, $(z \in X$ implies $z \in \mathscr{A})$" is a ppf in which *X* is the only free variable (z is bound and $\mathscr{A}$ is a constant class).   Consequently, by Axiom 2,

$$\{X \mid \text{for all sets } z, \ (z \in X \text{ implies } z \in \mathscr{A})\}$$

is a class whose members are precisely those sets which are subclasses of $\mathscr{A}$.

$\mathscr{P}(\mathscr{A})$ is called the **power class** of $\mathscr{A}$. Do not make the mistake of assuming that $\mathscr{P}(\mathscr{A})$ is the class of all *subclasses* of $\mathscr{A}$. The next proposition shows why not.

2.10   *Proposition*

If $\mathscr{A}$ is a proper class, then there does not exist a class whose elements are exactly the subclasses of $\mathscr{A}$.

2.11   **Axiom 3**
If X is a set, then $\mathscr{P}(X)$ is a set.

Axiom 3 is called the **power set axiom.**

2.12   **Proposition**
    (a) If $\mathscr{A}$ is a class, then there exists a class (denoted by) $\bigcup \mathscr{A}$ such that $X \in \bigcup \mathscr{A}$ iff there exists some $Y \in \mathscr{A}$ with $X \in Y$.
    (b) If $\mathscr{A}$ and $\mathscr{B}$ are classes, then there exists a class (denoted by) $\mathscr{A} \cup \mathscr{B}$ such that for all sets X, $X \in \mathscr{A} \cup \mathscr{B}$ iff $X \in \mathscr{A}$ or $X \in \mathscr{B}$.

The class $\bigcup \mathscr{A}$ is called the **union** of $\mathscr{A}$; $\mathscr{A} \cup \mathscr{B}$ is read $\mathscr{A}$ **union** $\mathscr{B}$.

2.13   **Axiom 4**
If X is a set, then $\bigcup X$ is a set.

Axiom 4 is called the **union axiom.**

2.14   **Proposition**
If U and V are sets, then there exists a class (denoted by) $\{U, V\}$ such that for all sets W, $W \in \{U, V\}$ iff $W = U$ or $W = V$.

2.15   **Axiom 5**
If U and V are sets, then so is $\{U, V\}$.

Axiom 5 is called the **axiom of unordered pairs.** The set $\{U, V\}$ is called the **unordered pair of U and V.** If $U = V$, then $\{U, V\}$ is written $\{U\}$ (or $\{V\}$, they're equal, by the axiom of extensionality) and is called **singleton** U.

2.16   **Proposition**
    (a) If $\mathscr{A}$ is a class, then there exists a class (denoted by) $\bigcap \mathscr{A}$ such that $X \in \bigcap \mathscr{A}$ iff for all $Y \in \mathscr{A}$, $X \in Y$.
    (b) If $\mathscr{A}$ and $\mathscr{B}$ are classes, then there exists a class (denoted by) $\mathscr{A} \cap \mathscr{B}$ such that for all sets X, $X \in \mathscr{A} \cap \mathscr{B}$ iff $X \in \mathscr{A}$ and $X \in \mathscr{B}$.

The class $\bigcap \mathscr{A}$ is called the **intersection** of $\mathscr{A}$; $\mathscr{A} \cap \mathscr{B}$ is read $\mathscr{A}$ **intersect** $\mathscr{B}$.

2.17   **Proposition**
There exists a unique class (denoted by) $\varnothing$ which has no elements.

The class $\varnothing$ is called the **empty** (or **null** or **void**) class.

2.18   Proposition

Let $\mathscr{A}$ and $\mathscr{B}$ be classes. There exists a class (denoted by) $\mathscr{A} - \mathscr{B}$ such that for all sets X, $X \in \mathscr{A} - \mathscr{B}$ iff $X \in \mathscr{A}$ and $X \notin \mathscr{B}$.

The class $\mathscr{A} - \mathscr{B}$ is called the **complement of** $\mathscr{B}$ **in** $\mathscr{A}$.

2.19   Proposition

If X and Y are sets, and if (X, Y) denotes the set $\{\{X\}, \{X, Y\}\}$, then

$$X = \bigcup\bigcap (X, Y),$$
$$Y = [\bigcap\bigcup(X, Y)] \cup [\bigcup\bigcup(X, Y) - \bigcup\bigcap(X, Y)].$$

Hence for all sets X, Y, X', and Y', $(X, Y) = (X', Y')$ iff $X = X'$ and $Y = Y'$.

The set $(X, Y) = \{\{X\}, \{X, Y\}\}$ is called the **ordered pair** of sets with **first element** X and **second element** Y.

2.20   Proposition

If $\mathscr{A}$ and $\mathscr{B}$ are classes, then there exists a class (denoted by) $\mathscr{A} \times \mathscr{B}$ such that $U \in \mathscr{A} \times \mathscr{B}$ iff there exist $X \in \mathscr{A}$ and $Y \in \mathscr{B}$ such that $U = (X, Y)$.

The class $\mathscr{A} \times \mathscr{B}$ is called the **cartesian product of** $\mathscr{A}$ **and** $\mathscr{B}$.

2.21   Definition

Let $\mathscr{A}$ and $\mathscr{B}$ be classes. A **function from** $\mathscr{A}$ **to** $\mathscr{B}$ is a subclass $\mathscr{F}$ of $\mathscr{A} \times \mathscr{B}$ such that for each $X \in \mathscr{A}$ there exists a unique $Y \in \mathscr{B}$ with $(X, Y) \in \mathscr{F}$; this Y is called the **value of** $\mathscr{F}$ **at** X and is denoted by $\mathscr{F}(X)$. We shall write $\mathscr{F} : \mathscr{A} \to \mathscr{B}$ to mean that $\mathscr{F}$ is a function from $\mathscr{A}$ to $\mathscr{B}$. A class $\mathscr{F}$ is a **function** iff there exist classes $\mathscr{A}$ and $\mathscr{B}$ such that $\mathscr{F} : \mathscr{A} \to \mathscr{B}$.*

2.22   Axiom 6

If $\mathscr{F}$ is a function and Z is a set, then the class†

$$\mathscr{F}[Z] = \{Y \,|\, \text{there exists } X \in Z \text{ such that } (X, Y) \in \mathscr{F}\}$$

is a set.

Axiom 6 is called the **axiom of replacement**.

---

* Observe that the definition of "function" (as opposed to "function from $\mathscr{A}$ to $\mathscr{B}$") is given above in terms of a non-ppf. In order that defined terms may be used in connection with the axiom of class formation, this situation should be avoided as much as possible. In this case you can see that the following definition of function (given by a ppf) is equivalent to the one above.
    Let $\mathscr{F}$ be a class. Then $\mathscr{F}$ is a **function** iff
    (a) for all $U \in \mathscr{F}$, there exist sets X and Y such that $U = (X, Y)$; and
    (b) if $(X, Y) \in \mathscr{F}$ and $(X, Z) \in \mathscr{F}$, then $Y = Z$.
† It is easy to verify using the axiom of class formation that $\mathscr{F}[Z]$ is a class. Such routine verifications will no longer be mentioned in the body of the text. They are presented in tabular form in Appendix C.

2.23   Proposition
If X is a set and $\mathscr{Y} \subseteq X$, then $\mathscr{Y}$ is a set.

2.24   Proposition
If X and Y are sets, then $X \times Y$ is a set.

2.25   Proposition
(a) If $\mathscr{A} \neq \varnothing$, then $\bigcap \mathscr{A}$ is a set.
(b) $\bigcap \varnothing = \{X \mid X$ is a set$\}$.

2.26   Definition
$\bigcap \varnothing$ is called the **universe** or **universal class**, and is denoted by $\mathfrak{U}$.

2.27   Proposition
(a) $\mathfrak{U} = \{X \mid X = X\}$.
(b) $\mathfrak{U}$ is not a set.

2.28   Axiom 7
If $\mathscr{A}$ is a nonempty class, then there exists a set X such that $X \in \mathscr{A}$ and $X \cap \mathscr{A} = \varnothing$.

Axiom 7 is called the **axiom of regularity**.

2.29   Proposition
If X and Y are sets, then
(a) $X \notin X$.
(b) If $X \in Y$, then $Y \notin X$.

Actually, there can be no "finite membership loops" $x_1 \in x_2 \in \cdots \in$ $\in x_n \in x_1$, or "infinitely descending membership chains" $\cdots \in y_n \in \cdots \in y_2 \in$ $\in y_1$. ("Finite" and "infinite" are defined in 2.132.)

2.30   Axiom 8
There exists a set X such that $\varnothing \in X$ and if $Y \in X$, then $Y \cup \{Y\} \in X$.

Axiom 8 is called the **axiom of infinity**.
Up to this point, we did not know that there were *any sets at all.* Indeed, if we take $\mathfrak{U} = \varnothing$, then we have a model for Axioms 1 through 7 in which there are no sets, and the only class is $\varnothing$. This also shows that we cannot know before Axiom 8 that $\varnothing$ is a set. This axiom effectively guarantees the existence of an infinite set and allows us (in 2.101) to define the set of natural numbers in our set theory.

2.31

If we are given a finite number of nonempty sets, say x_1, x_2, and x_3, then we can construct a function that "chooses" one element from each set as follows: $x_1 \neq \varnothing$, so there exists a $y_1 \in x_1$; $x_2 \neq \varnothing$, so there exists a $y_2 \in x_2$; $x_3 \neq \varnothing$, so there exists a $y_3 \in x_3$. By the axiom of class formation,

$$\mathscr{F} = \{Z \mid Z = (x_1, y_1) \text{ or } Z = (x_2, y_2) \text{ or } Z = (x_3, y_3)\}$$

is a class. Clearly $\mathscr{F} : \{x_1, x_2, x_3\} \to \bigcup \{x_1, x_2, x_3\}$ such that $\mathscr{F}(x_1) \in x_1$, $\mathscr{F}(x_2) \in x_2$, and $\mathscr{F}(x_3) \in x_3$. This works because we are able to put together a *finite number* of statements of the form $Z = (x_i, y_i)$ to build a ppf, and then apply Axiom 2 to obtain the desired "choice function".

This is a subtle point. As set-theoretic precision became more necessary, mathematicians came to realize that such an argument probably could not be used to define a choice function on just any class of sets. In 1940 Kurt Gödel proved that if the axioms already presented are consistent, then one can safely add a new axiom which says that such a choice *is* always possible. Finally, in 1963, Paul Cohen presented a model for set theory in which there was not always such a choice function. Thus the assumption of the existence of a choice function is independent of the present axioms. Since little (apparently) is to be gained by not assuming it and (as it turns out) a lot is to be gained by assuming it, we enlist this assumption as our next axiom.

Axiom 9
If X is a set of nonempty sets, then there is a function $C : X \to \bigcup X$ such that for every $Y \in X$, $C(Y) \in Y$.

Axiom 9 is called the **axiom of choice.*** The function C is called a **choice function.** Later we shall present a number of statements equivalent to the axiom of choice. For that reason, until 2.70, no proofs require Axiom 9, and none should use it.

This concludes our complete list of axioms.

†☐ **2.32**   **Proposition**
If $\mathscr{A}$ and $\mathscr{B}$ are subclasses of a class $\mathscr{C}$, then the following statements are equivalent.
 (a) $\mathscr{A} \subseteq \mathscr{B}$.
 (b) $\mathscr{C} - \mathscr{A} \supseteq \mathscr{C} - \mathscr{B}$.

---

* There is a stronger form of the axiom of choice which is also independent of the other axioms (Gödel, 1940; Cohen, 1963):
    AC2. There is a function $\mathscr{C} : \mathfrak{U} \to \mathfrak{U}$ such that for every nonempty set Y, $\mathscr{C}(Y) \in Y$.
    In most of mathematics, the "set form", Axiom 9, suffices (see, for example, 2.70). However, the "class form", AC2, is used occasionally, for example, in category theory.
† The symbol ☐ preceding the statement of a proposition means that similar results appeared in Chapter 1 and essentially the same proof works here.

(c) $\mathscr{B} = \mathscr{A} \cup \mathscr{B}$.

(d) $\mathscr{A} = \mathscr{A} \cap \mathscr{B}$.

☐ **2.33   Proposition**

Let $\mathscr{A}$, $\mathscr{B}$, and $\mathscr{C}$ be classes and let X be a set.   Then

(a) $\mathscr{A} = \mathscr{A} \cup \mathscr{A} = \mathscr{A} \cap \mathscr{A} = \mathscr{A} \cap (\mathscr{B} \cup \mathscr{A}) = \mathscr{A} \cup (\mathscr{B} \cap \mathscr{A})$.

(b) $\mathscr{B} - (\mathscr{B} - \mathscr{A}) = \mathscr{B} \cap \mathscr{A}$.

(c) $\mathscr{A} \cup \mathscr{B} = \mathscr{B} \cup \mathscr{A}$.

(d) $\mathscr{A} \cup (\mathscr{B} \cup \mathscr{C}) = (\mathscr{A} \cup \mathscr{B}) \cup \mathscr{C}$.

(e) $\mathscr{A} \cap (\mathscr{B} \cap \mathscr{C}) = (\mathscr{A} \cap \mathscr{B}) \cap \mathscr{C}$.

(f) $\mathscr{A} \cap (\bigcup \mathscr{C}) = \bigcup \{\mathscr{A} \cap Y \mid Y \in \mathscr{C}\}$.

(g) $X \cup (\bigcap \mathscr{C}) = \bigcap \{X \cup Y \mid Y \in \mathscr{C}\}$.

(h) $X - \bigcup \mathscr{C} = \bigcap \{X - Y \mid Y \in \mathscr{C}\}$ if $\mathscr{C} \neq \varnothing$.

(i) $X - \bigcap \mathscr{C} = \bigcup \{X - Y \mid Y \in \mathscr{C}\}$.

☐ **2.34   Proposition**

For any classes $\mathscr{A}$, $\mathscr{B}$, and $\mathscr{C}$,

$$\mathscr{A} \times (\mathscr{B} - \mathscr{C}) = (\mathscr{A} \times \mathscr{B}) - (\mathscr{A} \times \mathscr{C}).$$

☐ **2.35   Proposition**

For any classes $\mathscr{A}$ and $\mathscr{B}$,

(a) $(\bigcup \mathscr{A}) \times (\bigcup \mathscr{B}) = \bigcup \{X \times Y \mid (X, Y) \in \mathscr{A} \times \mathscr{B}\}$.

(b) $(\bigcap \mathscr{A}) \times (\bigcap \mathscr{B}) = \bigcap \{X \times Y \mid (X, Y) \in \mathscr{A} \times \mathscr{B}\}$, if $\mathscr{A}$ and $\mathscr{B}$ are not empty.

☐ **2.36   Corollary**

For any classes $\mathscr{A}$, $\mathscr{B}$, $\mathscr{C}$, and $\mathscr{D}$,

(a) $\mathscr{A} \times (\mathscr{B} \cup \mathscr{C}) = (\mathscr{A} \times \mathscr{B}) \cup (\mathscr{A} \times \mathscr{C})$.

(b) $\mathscr{A} \times (\mathscr{B} \cap \mathscr{C}) = (\mathscr{A} \times \mathscr{B}) \cap (\mathscr{A} \times \mathscr{C})$.

2.37   Definition

(a) Let $\mathscr{A}$ and $\mathscr{B}$ be classes.   A **relation from** $\mathscr{A}$ **to** $\mathscr{B}$ is a subclass of $\mathscr{A} \times \mathscr{B}$.   A relation from $\mathscr{A}$ to $\mathscr{A}$ is called a **relation on** $\mathscr{A}$.   A class $\mathscr{R}$ is a **relation** provided that there are classes $\mathscr{A}$ and $\mathscr{B}$ such that $\mathscr{R}$ is a relation from $\mathscr{A}$ to $\mathscr{B}$.

(b) If $\mathscr{R}$ is a relation, then the **domain** of $\mathscr{R}$ is the class

$$\{X \mid \text{there exists a } Y \text{ such that } (X, Y) \in \mathscr{R}\}.$$

(c) The **inverse** of a relation $\mathscr{R}$ is the relation*

$$\mathscr{R}^{-1} = \{(X, Y) \mid (Y, X) \in \mathscr{R}\}.$$

---

* Note that $\{(X, Y) \mid (Y, X) \in \mathscr{R}\}$ is just a shorthand notation for $\{Z \mid \text{there exists a set } X$ such that (there exists a set Y such that $(Y, X) \in \mathscr{R}$ and $Z = (X, Y))\}$. We will often use such shorthand notations when the meaning is clear.

(d) If $\mathscr{R}$ is a relation and $\mathscr{C}$ is a class, then the **image of $\mathscr{C}$ under $\mathscr{R}$** is the class

$$\mathscr{R}[\mathscr{C}] = \{Y \mid \text{there exists } X \in \mathscr{C} \text{ such that } (X, Y) \in \mathscr{R}\}.$$

(e) If $\mathscr{R}$ is a relation and $\mathscr{C}$ is a class, then the **restriction of $\mathscr{R}$ to $\mathscr{C}$** is the relation

$$\mathscr{R} \mid \mathscr{C} = \mathscr{R} \cap (\mathscr{C} \times \mathscr{R}[\mathscr{C}]).$$

The **total restriction** of $\mathscr{R}$ to $\mathscr{C}$ is the relation $\mathscr{R} \cap (\mathscr{C} \times \mathscr{C})$.

(f) If $\mathscr{R}$ and $\mathscr{S}$ are relations, then the **composition of $\mathscr{R}$ with $\mathscr{S}$** (or **$\mathscr{R}$ followed by $\mathscr{S}$**) is the relation

$$\mathscr{S} \circ \mathscr{R} = \{(X, Z) \mid \text{there exists } Y \text{ such that } (X, Y) \in \mathscr{R} \text{ and } (Y, Z) \in \mathscr{S}\}.$$

(g) For any class $\mathscr{A}$, the **identity relation** on $\mathscr{A}$ is the class $1_{\mathscr{A}} = \{(X, X) \mid X \in \mathscr{A}\}$.

(h) If $\mathscr{F}$ and $\mathscr{G}$ are functions, and if $\mathscr{G}$ is a restriction of $\mathscr{F}$, then $\mathscr{F}$ is called an **extension of $\mathscr{G}$**.

□ **2.38   Proposition**

Let $\mathscr{R}$, $\mathscr{S}$, and $\mathscr{T}$ be relations, and $\mathscr{V}$ be a class of relations. Then

(a) If $\mathscr{P} \subseteq \mathscr{R}$ and $\mathscr{Q} \subseteq \mathscr{S}$, then $\mathscr{P} \circ \mathscr{Q} \subseteq \mathscr{R} \circ \mathscr{S}$.

(b) $(\mathscr{R} \circ \mathscr{S}) \circ \mathscr{T} = \mathscr{R} \circ (\mathscr{S} \circ \mathscr{T})$.

(c) $\mathscr{R} \circ (\mathscr{S} \cup \mathscr{T}) = (\mathscr{R} \circ \mathscr{S}) \cup (\mathscr{R} \circ \mathscr{T})$.

(d) $\mathscr{R} \circ (\mathscr{S} \cap \mathscr{T}) \subseteq (\mathscr{R} \circ \mathscr{S}) \cap (\mathscr{R} \circ \mathscr{T})$.

(e) $(\mathscr{R}^{-1})^{-1} = \mathscr{R}$.

(f) $(\mathscr{R} \circ \mathscr{S})^{-1} = \mathscr{S}^{-1} \circ \mathscr{R}^{-1}$.

(g) If $\mathscr{R}$ is a set, then $\mathscr{R} \circ \bigcup \mathscr{V} = \bigcup \{\mathscr{R} \circ U \mid U \in \mathscr{V}\}$.

(h) $\mathscr{R} \circ \bigcap \mathscr{V} \subseteq \bigcap \{\mathscr{R} \circ U \mid U \in \mathscr{V}\}$.

Note that part (g) is not true if we omit the restriction that $\mathscr{R}$ be a set. For example, if $\mathscr{R}$ is the universe, $\mathfrak{U}$, and V is any nonempty set, then $\mathscr{R} \circ V$ is a proper class, so $\mathscr{R} \circ V \notin \{\mathscr{R} \circ U \mid U \in \{V\}\} = \varnothing$. This is because $\{\mathscr{R} \circ U \mid U \in \{V\}\}$ means the class of all *sets* $\mathscr{R} \circ U$ for $U \in \{V\}$. This somewhat subtle point must also be taken into account in several later results (for example, 2.55).

2.39   Definition

(a) A function $\mathscr{F} : \mathscr{A} \to \mathscr{B}$ is **one-to-one** or is **injective** iff for all $X, X' \in \mathscr{A}$, $\mathscr{F}(X) = \mathscr{F}(X')$ implies $X = X'$.

(b) A function $\mathscr{F} : \mathscr{A} \to \mathscr{B}$ is **onto** $\mathscr{B}$ or is **surjective** iff $\mathscr{B} = \mathscr{F}[\mathscr{A}]$.

(c) A function $\mathscr{F} : \mathscr{A} \to \mathscr{B}$ is a **bijection** iff $\mathscr{F}$ is one-to-one and onto $\mathscr{B}$.

(d) A function $\mathscr{F} : \mathscr{A} \times \mathscr{A} \to \mathscr{A}$ is called a **binary operation on $\mathscr{A}$**. In this case $\mathscr{F}((X, Y))$ is usually written $X \mathscr{F} Y$.

☐ *2.40*   **Proposition**
Let $\mathscr{F}$ and $\mathscr{G}$ be functions, $\mathscr{R}$ be a relation, and $\mathscr{A}$ and $\mathscr{B}$ be classes.   Then
(a) $\mathscr{F} \circ \mathscr{G}$ is a function.
(b) $\mathscr{R}[\mathscr{A} \cup \mathscr{B}] = \mathscr{R}[\mathscr{A}] \cup \mathscr{R}[\mathscr{B}]$.
(c) $\mathscr{R}[\mathscr{A} \cap \mathscr{B}] \subseteq \mathscr{R}[\mathscr{A}] \cap \mathscr{R}[\mathscr{B}]$.
(d) $\mathscr{F}^{-1}[\mathscr{A} \cap \mathscr{B}] = \mathscr{F}^{-1}[\mathscr{A}] \cap \mathscr{F}^{-1}[\mathscr{B}]$.
(e) $\mathscr{F}^{-1}[\mathscr{A} - \mathscr{B}] = \mathscr{F}^{-1}[\mathscr{A}] - \mathscr{F}^{-1}[\mathscr{B}]$.
(f) If $\mathscr{R}$ is a set, then $\mathscr{R}[\bigcup \mathscr{A}] = \bigcup \{\mathscr{R}[X] \mid X \in \mathscr{A}\}$.
(g) $\mathscr{R}[\bigcap \mathscr{A}] \subseteq \bigcap \{\mathscr{R}[X] \mid X \in \mathscr{A}\}$.

☐ *2.41*   **Proposition**
Let $\mathscr{F} : \mathscr{A} \to \mathscr{B}$.   The following statements are equivalent.
(a) $\mathscr{F}$ is injective.
(b) $\mathscr{F}^{-1} \circ \mathscr{F} = 1_{\mathscr{A}}$.
(c) If $\mathscr{A} \neq \varnothing$, then there exists a function $\mathscr{G} : \mathscr{B} \to \mathscr{A}$ such that $\mathscr{G} \circ \mathscr{F} = 1_{\mathscr{A}}$.
(d) $\mathscr{F}$ is **left cancellable** (i.e., if $\mathscr{H} : \mathscr{C} \to \mathscr{A}$ and $\mathscr{K} : \mathscr{C} \to \mathscr{A}$, and if $\mathscr{F} \circ \mathscr{H} = \mathscr{F} \circ \mathscr{K}$, then $\mathscr{H} = \mathscr{K}$).

☐ *2.42*   **Proposition**
Let $\mathscr{F} : \mathscr{A} \to \mathscr{B}$.   The following statements are equivalent.
(a) $\mathscr{F}$ is surjective.
(b) $\mathscr{F} \circ \mathscr{F}^{-1} = 1_{\mathscr{B}}$.
(c) There exists a relation $\mathscr{G}$ from $\mathscr{B}$ to $\mathscr{A}$ such that $\mathscr{F} \circ \mathscr{G} = 1_{\mathscr{B}}$.
(d) $\mathscr{F}$ is **right cancellable** (i.e., if $\mathscr{H} : \mathscr{B} \to \mathscr{C}$ and $\mathscr{K} : \mathscr{B} \to \mathscr{C}$ and if $\mathscr{H} \circ \mathscr{F} = \mathscr{K} \circ \mathscr{F}$, then $\mathscr{H} = \mathscr{K}$).

If $\mathscr{A}$ is a set and if we were permitted the use of the axiom of choice, we could say "function" instead of "relation" in part (c) of this proposition.

☐ *2.43*   **Corollary**
A function $\mathscr{F} : \mathscr{A} \to \mathscr{B}$ is a bijection iff there exists a $\mathscr{G} : \mathscr{B} \to \mathscr{A}$ such that $\mathscr{F} \circ \mathscr{G} = 1_{\mathscr{B}}$ and $\mathscr{G} \circ \mathscr{F} = 1_{\mathscr{A}}$.

2.44   **Definition**
Let $\mathscr{R}$ be a relation on $\mathscr{A}$.   Then
(a) $\mathscr{R}$ is **reflexive on** a subclass $\mathscr{B}$ of $\mathscr{A}$ iff $\mathscr{R} \supseteq 1_{\mathscr{B}}$.
(b) $\mathscr{R}$ is **symmetric** iff $\mathscr{R} = \mathscr{R}^{-1}$.
(c) $\mathscr{R}$ is **transitive** iff $\mathscr{R} \circ \mathscr{R} \subseteq \mathscr{R}$.
(d) $\mathscr{R}$ is **antisymmetric** iff $\mathscr{R} \cap \mathscr{R}^{-1} \subseteq 1_{\mathscr{A}}$.

2.45   **Definition**
(a) An **equivalence relation on** $\mathscr{A}$ is a relation that is symmetric, transitive, and reflexive on $\mathscr{A}$.

(b) If $\mathscr{R}$ is an equivalence relation on $\mathscr{A}$ and $X \in \mathscr{A}$, then $\mathscr{R}[\{X\}]$ is called the **equivalence class of X modulo $\mathscr{R}$**.   When there is no possibility of confusion, this is sometimes abbreviated as $[X]$.

2.46   **Proposition**
Let $\mathscr{R}$ be an equivalence relation on $\mathscr{A}$.   Then, for all $X \in \mathscr{A}$,
    (a) $X \in \mathscr{R}[\{X\}]$.
    (b) $\mathscr{R}[\{X\}] \times \mathscr{R}[\{X\}] \subseteq \mathscr{R}$.
    (c) $Y \in \mathscr{R}[\{X\}]$ iff $\mathscr{R}[\{X\}] = \mathscr{R}[\{Y\}]$.

2.47   **Definition**
A **partition** of a set X is a set $Y \subseteq \mathscr{P}(X)$ such that
    (a) $\bigcup Y = X$,
    (b) $\varnothing \notin Y$, and
    (c) if $U, V \in Y$, then either $U = V$ or $U \cap V = \varnothing$.

☐ **2.48**   **Theorem**
A reflexive relation R on a set X is an equivalence relation on X iff $\{R[\{Y\}] \mid Y \in X\}$ is a partition of X.

2.49   **Definition**
Let R be an equivalence relation on a set X.
    (a) The **quotient set** X/R is the set $\{R[\{Y\}] \mid Y \in X\}$.
    (b) The **quotient** (or **canonical** or **natural**) **map** of X onto X/R is the function

$$\eta_R = \{(Y, R[\{Y\}]) \mid Y \in X\}.$$

Note that this definition is made only for sets.   In general one cannot form a quotient class of a *class* modulo a relation, since some of the equivalence classes may be proper classes, and these cannot be members of any class.   In certain very special cases, of course, it might be possible (when all equivalence classes are sets), but this must be decided on an individual basis.

2.50

On certain occasions, the difficulty outlined in the previous paragraph can be overcome by using the idea of a "system of representatives".   We are not claiming here that equivalence relations do have systems of representatives.   As it happens, such an assertion for sets is equivalent to the axiom of choice (see 2.68 and 2.69).

Definition
Let $\mathscr{R}$ be an equivalence relation on a class $\mathscr{A}$.   A **system of representatives** for $\mathscr{R}$ is a class $\mathscr{S} \subseteq \mathscr{A}$ such that for every $X \in \mathscr{A}$, $\mathscr{S} \cap \mathscr{R}[\{X\}]$ is a singleton.

2.51   **Proposition**
Let $\mathscr{G} : \mathscr{A} \to \mathscr{B}$.   Then

(a) $\{(X, Y) \mid \mathscr{G}(X) = \mathscr{G}(Y)\}$ is an equivalence relation on $\mathscr{A}$ (called the **kernel of** $\mathscr{G}$).

(b) if $\mathscr{A}$ is a set, then every equivalence relation on $\mathscr{A}$ is the kernel of some function.

2.52   **Proposition**
(a) The total restriction $\mathscr{R} \cap (\mathscr{B} \times \mathscr{B})$ of an equivalence relation $\mathscr{R}$ on $\mathscr{A}$ to a subclass $\mathscr{B}$ of $\mathscr{A}$ is an equivalence relation on $\mathscr{B}$.

(b) The intersection of any nonempty class of equivalence relations on $\mathscr{A}$ is an equivalence relation on $\mathscr{A}$.

2.53   **Definition**
A **family indexed by** a class $\mathscr{I}$ is a function $\mathscr{F} : \mathscr{I} \to \mathfrak{U}$.   If for each $i \in \mathscr{I}$, $\mathscr{F}(i)$ is denoted by A_i, then the family is denoted by $(A_i)_{i \in \mathscr{I}}$, $(A_i)_{\mathscr{I}}$, or (A_i), depending upon how much notation is needed.

2.54   **Definition**
Let $(A_i)_{\mathscr{I}}$ be a family of sets.

(a) The **cartesian product** of $(A_i)_{\mathscr{I}}$ is the class

$$\Pi(A_i) = \{F \mid F : \mathscr{I} \to \bigcup\{A_i \mid i \in \mathscr{I}\} \text{ such that } F(i) \in A_i \text{ for all } i \in \mathscr{I}\}.$$

This is sometimes denoted more briefly by ΠA_i.

(b) For $j \in \mathscr{I}$, the jth **projection function** is the function

$$\pi_j = \{(F, F(j)) \mid F \in \Pi(A_i)\}.$$

2.55   **Proposition**
$\Pi(A_i)$ is always a set.

2.56   **Theorem**
Let $(A_i)_I$ be a family of sets indexed by a set.

(a) The cartesian product ΠA_i together with the projection functions $\pi_i : \Pi A_i \to A_i$ is "universal" in the following sense: For every set Y together with functions $f_i : Y \to A_i$ for all $i \in I$, there exists a unique function $F : Y \to \Pi A_i$ such that for each $i \in I$ the diagram

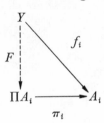

commutes.

(b) The above universality characterizes the cartesian product; i.e., if Z is a set together with functions $\rho_i : Z \to A_i$ for all $i \in I$ such that for every set Y together with functions $f_i : Y \to A_i$ for all $i \in I$ there exists a unique function $F : Y \to Z$ such that for all $i \in I$, the diagram

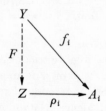

commutes; then there exists a bijection $H : Z \to \Pi A_i$ such that for all $i \in I$, the diagram

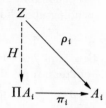

commutes.

2.57   Definition

(a) The **disjoint union** of a family $(A_i)_{\mathscr{I}}$ of sets is the class

$$\sum (A_i) = \bigcup \{A_i \times \{i\} \mid i \in \mathscr{I}\}.$$

This is sometimes denoted more briefly by $\sum A_i$.

(b) For each $j \in \mathscr{I}$ the function

$$\mu_j = \{(X, (X, j)) \mid X \in A_j\}$$

from A_j to $\sum A_i$ is called the jth **injection function.**

(c) If $\mathscr{A}$ and $\mathscr{B}$ are classes, the **disjoint union of $\mathscr{A}$ and $\mathscr{B}$** is the class

$$\mathscr{A} \uplus \mathscr{B} = (\mathscr{A} \times \{\varnothing\}) \cup (\mathscr{B} \times \{\{\varnothing\}\}).$$

□ **2.58   Theorem**

Let $(A_i)_{\mathscr{I}}$ be a family of sets.

(a) The disjoint union $\sum A_i$ together with the injections $\mu_i : A_i \to \sum A_i$ is "universal" in the following sense: For every class $\mathscr{Y}$ together with functions $f_i : A_i \to \mathscr{Y}$ for all $i \in \mathscr{I}$, there exists a unique function

$\mathcal{F} : \sum A_i \to \mathcal{Y}$ such that for each $i \in \mathcal{I}$ the diagram

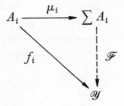

commutes.

(b) The above universality characterizes the disjoint union; i.e., if $\mathcal{Z}$ is a class together with functions $\delta_i : A_i \to \mathcal{Z}$ for each $i \in \mathcal{I}$, such that for every class $\mathcal{Y}$ together with functions $f_i : A_i \to \mathcal{Y}$ for each $i \in \mathcal{I}$, there exists a unique function $\mathcal{F} : \mathcal{Z} \to \mathcal{Y}$ such that for each $i \in \mathcal{I}$, the diagram

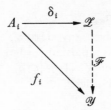

commutes; then there exists a bijection $\mathcal{H} : \sum A_i \to \mathcal{Z}$ such that for each $i \in \mathcal{I}$, the diagram

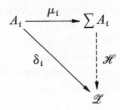

commutes.

2.59   *Definition*

(a) A relation $\mathcal{R}$ on a class $\mathcal{A}$ is a **quasi-order** on $\mathcal{A}$ (and $(\mathcal{A}, \mathcal{R})$* is a **quasi-ordered class**) iff $\mathcal{R}$ is transitive and reflexive on $\mathcal{A}$.

---

* We have not defined "ordered pair" for classes. One could think of a quasi-ordered class as a "class together with a quasi-order relation", or define an ordered pair of classes, for example, to be their disjoint union—the only thing important about ordered pairs being that two ordered pairs are equal iff their first and second coordinates are equal. A third approach is to define a quasi-ordered class to be a quasi-order relation; since quasi-orders are reflexive, the underlying class can be recovered at will.

(b) An antisymmetric quasi-order $\mathscr{R}$ on a class $\mathscr{A}$ is called a **partial order** on $\mathscr{A}$ (and $(\mathscr{A}, \mathscr{R})$ is called a **partially ordered class**).

(c) An **ideal** of a quasi-ordered class $(\mathscr{A}, \mathscr{R})$ is a subclass $\mathscr{I}$ of $\mathscr{A}$ such that whenever $X \in \mathscr{I}$ and $Y \in \mathscr{A}$ with $(Y, X) \in \mathscr{R}$, then $Y \in \mathscr{I}$.

(d) An **(order) isomorphism** between quasi-ordered classes $(\mathscr{A}, \mathscr{R})$ and $(\mathscr{A}', \mathscr{R}')$ is a bijection $\mathscr{F} : \mathscr{A} \to \mathscr{A}'$ such that

$$(X, Y) \in \mathscr{R} \text{ iff } (\mathscr{F}(X), \mathscr{F}(Y)) \in \mathscr{R}'.$$

If there is such an isomorphism, then $(\mathscr{A}, \mathscr{R})$ and $(\mathscr{A}', \mathscr{R}')$ are said to be **(order) isomorphic**.

The usual symbol for a quasi-order relation is $\leq$ and one usually writes $X \leq Y$ instead of $(X, Y) \in \leq$. We write $X < Y$ to mean $X \leq Y$ and $X \neq Y$; and $X > Y$ to mean $Y < X$.

2.60   Examples

(a) Given any class $\mathscr{A}$, $(\mathscr{P}(\mathscr{A}), \subseteq)$ is a partially ordered class; indeed $(\mathfrak{U}, \subseteq)$ is a partially ordered class.

(b) Let $X = \{\varnothing, \{\varnothing\}\}$. Then $X \times X$ is a quasi-order on X that is not a partial order.

2.61   Examples

These and Examples 2.66 and 2.133 are included to illustrate definitions and are not necessary for the development of set theory. The sets involved are developed fully in Chapter 3. Let $\mathbf{N}$, $\mathbf{Z}$, $\mathbf{Q}$, and $\mathbf{R}$ be, respectively, the sets of positive integers, all integers, rational numbers, and real numbers.

(a) The usual ordering on $\mathbf{R}$ is a partial order.

(b) Each nonempty proper ideal of $\mathbf{R}$ is of either the form $\{x \mid x < a\}$ or the form $\{x \mid x \leq a\}$ for some $a \in \mathbf{R}$.

(c) No two of the sets $\mathbf{N}$, $\mathbf{Z}$, $\mathbf{Q}$, and $\mathbf{R}$ with the usual order are isomorphic.

2.62   Proposition

Let $(\mathscr{A}, \leq)$ be a quasi-ordered class. If $\mathscr{B}$ is a nonempty class of ideals of $\mathscr{A}$, then $\bigcup \mathscr{B}$ and $\bigcap \mathscr{B}$ are ideals of $\mathscr{A}$.

2.63   Proposition

If X is a set and $\leq$ is a quasi-order relation on X, then

(a) $R = \{(x, y) \mid x, y \in X \text{ and } x \leq y \text{ and } y \leq x\}$ is an equivalence relation on X.

(b) X/R with the relation $\{(R[\{x\}], R[\{y\}]) \mid x \leq y\}$ is a partially ordered set.

2.64   **Definition**

Let $(\mathscr{A}, \leq)$ be a partially ordered class.

(a) A subclass $\mathscr{B}$ of $\mathscr{A}$ is **totally ordered** by (the total restriction of) $\leq$ (and $(\mathscr{B}, \leq \cap (\mathscr{B} \times \mathscr{B}))$ is called a **totally ordered class**) provided that for all $X, Y \in \mathscr{B}$ either $X \leq Y$ or $Y \leq X$.

(b) $X \in \mathscr{A}$ is an **upper bound** for a subclass $\mathscr{B}$ of $\mathscr{A}$ provided that $Y \leq X$ for all $Y \in \mathscr{B}$. **Lower bound** is defined similarly. $\mathscr{B}$ is said to be **bounded** iff it has an upper bound and a lower bound.

(c) $X \in \mathscr{A}$ is a **maximal element** of $\mathscr{A}$ provided that $Z \in \mathscr{A}$ and $X \leq Z$ together imply that $X = Z$ (i.e., provided that no element of $\mathscr{A}$ exceeds X). **Minimal element** is defined similarly.

(d) $X \in \mathscr{A}$ is the **greatest element** of $\mathscr{A}$ provided that X is an upper bound for $\mathscr{A}$. **Least element** is defined similarly.

(e) $\leq$ is called a **well-order** on $\mathscr{A}$ (and $(\mathscr{A}, \leq)$ is called a **well-ordered class**) provided that every nonempty subset of $\mathscr{A}$ has a least element.

(f) $X \in \mathscr{A}$ is the **supremum** (or **least upper bound**) of a subclass $\mathscr{B}$ of $\mathscr{A}$ provided that X is the least element of the class of all upper bounds of $\mathscr{B}$. In this case we write $X = \sup \mathscr{B}$. Similarly **infimum** is synonymous with **greatest lower bound**, and the corresponding notation is $\inf \mathscr{B}$.

2.65   **Examples**

(a) For any set X, the greatest element of $\mathscr{P}(X)$ is X; the least element is $\varnothing$; an upper bound (in fact, the supremum) of any subset Y of $\mathscr{P}(X)$ is $\bigcup Y$ and a lower bound (in fact, the infimum) is $X \cap (\bigcap Y)$. In general there are other upper and lower bounds. If X has more than one member, then $\mathscr{P}(X)$ is not totally ordered.

(b) The set

$$\{\varnothing, \{\varnothing\}, \{\varnothing, \{\varnothing\}\}, \{\varnothing, \{\varnothing\}, \{\varnothing, \{\varnothing\}\}\}\}$$

is well-ordered by $\subseteq$.

(c) If A and B are nonempty disjoint sets, then $(\mathscr{P}(A) \cup \mathscr{P}(B), \subseteq)$ has two maximal elements but no greatest element. It has one minimal element which is also the least element.

2.66   **Examples** (cf. 2.61)

(a) **N** with the usual order is well-ordered, but **Z**, **Q**, and **R** are not well-ordered.

(b) The set $\mathbf{N} \cup \{\infty\}$ with the relation

$$\{(a, b) \mid \text{either } b = \infty \text{ or } a \leq b \text{ in } \mathbf{N}\}$$

is well-ordered and each of its subsets has a supremum, but not necessarily a greatest element.

2.67   *Proposition*
Every well-ordered class is totally ordered.

2.68

We now consider some statements that are equivalent to the axiom of choice (see 2.31).

Zorn's Lemma
If X is a nonempty partially ordered set every totally ordered subset of which has an upper bound, then X has a maximal element.

Hausdorff Maximality Principle
If X is a partially ordered set and T is a totally ordered subset of X, then there exists a maximal (with respect to $\subseteq$) totally ordered subset of X containing T.

Theorem
The following statements are equivalent (assuming Axioms 1 through 8).
    (a) Hausdorff Maximality Principle.
    (b) Zorn's Lemma.
    (c) Every equivalence relation on a *set* has a system of representatives.*
    (d) Axiom of choice.
    (e) For any set X, there exists a relation $\leq$ such that $(X, \leq)$ is a well-ordered set.

2.69

The following statements are equivalent to the "class form" of the axiom of choice, AC2 (see 2.31).
    (a) Every class can be well-ordered so that each ideal is a set.
    (b) Every equivalence relation on a class has a system of representatives.

The importance of part (b) becomes clear when we realize that we cannot always form the class of equivalence classes of an equivalence relation (see 2.49 and 2.50) but we can consider a system of representatives.

The proofs of (a) $\Rightarrow$ (b) and (b) $\Rightarrow$ AC2 are not difficult. The proof of AC2 $\Rightarrow$ (a) requires a good amount of work and machinery beyond the scope of this presentation.

From now on, we assume the axiom of choice (Axiom 9) and its equivalent formulations.

---

* Recall that this was a theorem in intuitive set theory (cf. 1.52).

2.70   **Examples**

The following propositions are typical applications of the axiom of choice.

(a) *Proposition.*   If R is any relation on a set X, then there is a maximal subset Z of X for which $Z \times Z \subseteq R$.

PROOF   (The following is a "typical Zorn's Lemma argument".)

Let $S = \{Y \subseteq X \mid Y \times Y \subseteq R\}$, partially ordered by $\subseteq$.   Since $\varnothing \in S$, S is not empty.   Let T be a totally ordered subset of S.   We will show that $\bigcup T$ is an upper bound for T in $(S, \subseteq)$.   Let $(a, b) \in (\bigcup T) \times (\bigcup T)$. Then there exist $Y_1, Y_2 \in T$ such that $a \in Y_1$ and $b \in Y_2$.   Since T is totally ordered, either $Y_1 \subseteq Y_2$ or $Y_2 \subseteq Y_1$; let W be the larger of the two.   Then $(a, b) \in W \times W \subseteq R$.   Hence $\bigcup T \in S$.   Clearly $\bigcup T$ contains each element of T, so $\bigcup T$ is an upper bound for T.   By Zorn's Lemma, then, S has a maximal element; and this is Z.

(b) *Proposition.*   If $(X_i)_I$ is a family of nonempty sets indexed by a set I, then $\Pi(X_i)_I$ is nonempty and each projection function is surjective.

(c) *Proposition.*   Let X be a nonempty partially ordered set every totally ordered subset of which has an upper bound.   If $f : X \to X$ such that $f(y) \geq y$ for all $y \in X$, then there exists some $y_0 \in X$ for which $f(y_0) = y_0$.

(d) *Proposition.*   Every partial order on a set is contained in a total order.

(e) *Proposition.**   Every ideal in a ring with identity is contained in a maximal ideal.

(f) *Proposition.**   Every field has an algebraic closure.

(g) *Proposition.**   Every filter on a set is contained in a maximal filter (see 4.125).

(h) *Proposition.**   Every ideal in a boolean algebra is contained in a maximal ideal.

(i) *Proposition.**   Any product of compact topological spaces is compact (see 4.158).

2.71   **Definition**

Let $(\mathscr{A}, \leq)$ be a totally ordered class.   For each $X \in \mathscr{A}$, the class

$$\mathscr{S}_X = \{Y \in \mathscr{A} \mid Y < X\}$$

is called the **initial segment** of $\mathscr{A}$ determined by X.

2.72   **Proposition**

If $(\mathscr{A}, \leq)$ is a well-ordered class every initial segment of which is a set, then every nonempty subclass of $\mathscr{A}$ has a least element.

---

* Some of the terms in this proposition have not been defined here.   The proposition is for illustration and need not be proved now.

2.73   **Theorem (Principle of Transfinite Induction)**
Let $(\mathscr{A}, \leq)$ be a well-ordered class every initial segment of which is a set, and let $\mathscr{B} \subseteq \mathscr{A}$. Suppose that for each $X \in \mathscr{A}$, $\mathscr{S}_X \subseteq \mathscr{B}$ implies $X \in \mathscr{B}$. Then $\mathscr{B} = \mathscr{A}$.

2.74   **Definition**
Let $(\mathscr{A}, \leq)$ be a well-ordered class and let $X \in \mathscr{A}$. The **immediate successor** of X is the least element of the class $\{Y \in \mathscr{A} \mid Y > X\}$. The **immediate predecessor** of X is the greatest element of the class $\{Y \in \mathscr{A} \mid Y < X\}$. If the indicated class has no least (resp., greatest) element, then X has no immediate successor (predecessor).

2.75   **Proposition**
Let $(\mathscr{A}, \leq)$ be a well-ordered class.
   (a) If each initial segment of $\mathscr{A}$ is a set and $X \in \mathscr{A}$ is not the greatest element of $\mathscr{A}$, then X has an immediate successor.
   (b) If Y is the immediate successor of X, then X is the immediate predecessor of Y.

2.76   **Theorem**
Let $(X, \leq)$ be a well-ordered set and let $f : X \to X$ be such that $f(a) < f(b)$ whenever $a < b$. Then $x \leq f(x)$ for all $x \in X$.

2.77   **Corollary**
There exists *at most* one isomorphism between any two well-ordered sets.

2.78   **Corollary**
No initial segment of a well-ordered set $(X, \leq)$ is isomorphic to $(X, \leq)$.

2.79   **Proposition**
Let $(\mathscr{A}, \leq)$ be a well-ordered class.
   (a) Every subclass of $\mathscr{A}$ is well-ordered by (the total restriction of) $\leq$.
   (b) If every initial segment of $\mathscr{A}$ is a set, then the initial segments are exactly the proper ideals of $\mathscr{A}$ (i.e., the ideals not equal to $\mathscr{A}$).

2.80   **Proposition**
If $(X, \leq)$ is a well-ordered set and $(\mathscr{Y}, \leq)$ is a well-ordered class every initial segment of which is a set, then there is at most one ideal $\mathscr{I}$ of $\mathscr{Y}$ that is isomorphic to X, and at most one isomorphism from X to $\mathscr{I}$.

2.81   **Theorem**
If $(X, \leq)$ is a well-ordered set and $(\mathscr{Y}, \leq)$ is a well-ordered class every initial segment of which is a set, then exactly one of the following statements is true:
   (a) X is isomorphic to $\mathscr{Y}$.

(b) X is isomorphic to an initial segment of $\mathscr{Y}$.

(c) $\mathscr{Y}$ is isomorphic to an initial segment of X.

Moreover, in each case, the isomorphism is unique.

2.82   Proposition

Any subset of a well-ordered set $(X, \leq)$ is isomorphic either to $(X, \leq)$ or to an initial segment of $(X, \leq)$.

2.83   Definition

Two well-ordered sets $(X, \leq)$ and $(X', \leq')$ are said to be **similar** (written $X \simeq X'$) provided that there is an isomorphism between them.

2.84   Theorem

$\simeq$ is an equivalence relation on the class of all well-ordered sets.

2.85

We proceed to a construction of the ordinal numbers. Our interest in them lies in the fact that, as we shall see, the class of all ordinal numbers forms a system of representatives for the equivalence relation $\simeq$. Even though we cannot gather the equivalence classes of $\simeq$ together into a class, we can (and do) consider the class of all ordinal numbers.

2.86   Definition

An **ordinal** is a class $\mathscr{A}$ satisfying the following two conditions:

(a) If $X \in \mathscr{A}$ and $Y \in \mathscr{A}$, then at least one of the following holds: $X \in Y$, $X = Y$, or $Y \in X$.

(b) Every element of $\mathscr{A}$ is a subset of $\mathscr{A}$.

An **ordinal number** is an ordinal that is a set.

In view of Proposition 2.29, "at least" in the above definition can be replaced by "exactly".

2.87   Proposition

Every element of an ordinal is an ordinal number.

2.88   Proposition

If $\mathscr{A}$ and $\mathscr{B}$ are ordinals and $\mathscr{A} \subsetneq \mathscr{B}$, then $\mathscr{A} \in \mathscr{B}$.

2.89   Proposition

If $\mathscr{A}$ and $\mathscr{B}$ are ordinals, then either $\mathscr{A} \subseteq \mathscr{B}$ or $\mathscr{B} \subseteq \mathscr{A}$.

2.90   Corollary

If $\mathscr{A}$ and $\mathscr{B}$ are ordinals, then exactly one of the following holds: $\mathscr{A} \in \mathscr{B}$, $\mathscr{A} = \mathscr{B}$, or $\mathscr{B} \in \mathscr{A}$.

2.91   Theorem
The class $\mathfrak{O}$ of all ordinal numbers is an ordinal that is not an ordinal number and it is the only ordinal that is not an ordinal number.

2.92   Theorem
(a) $\mathfrak{O}$ is well-ordered by $\subseteq$.
(b) Every initial segment of $(\mathfrak{O}, \subseteq)$ is an ordinal number and every ordinal number is an initial segment of $(\mathfrak{O}, \subseteq)$.

2.93   Corollary
Every ordinal number is well-ordered by $\subseteq$.

2.94   Proposition
(a) If A is an ordinal number, then the immediate successor of A in $(\mathfrak{O}, \subseteq)$ is $A^+ = A \cup \{A\}$.
(b) If $\mathscr{A}$ is a class of ordinal numbers, then $\bigcup \mathscr{A}$ is an ordinal. Hence every set of ordinal numbers has a supremum in $(\mathfrak{O}, \subseteq)$.

2.95   Theorem
Two ordinals are similar iff they are equal.

2.96   Theorem
$\mathfrak{O}$ is a system of representatives for $\simeq$ on the class of all well-ordered sets.

2.97   Definition
If $(X, \leq)$ is any well-ordered set, then the unique ordinal number A for which $(A, \subseteq) \simeq (X, \leq)$ is denoted by $\mathrm{ord}(X, \leq)$, and is called the **ordinal number of** $(X, \leq)$.

2.98   Proposition
If $A \subseteq B \in \mathfrak{O}$, then $\mathrm{ord}(A, \subseteq) \subseteq B$.

2.99   Definition
A **successor set** is a set A for which
(a) $\varnothing \in A$, and
(b) if $a \in A$, then $a^+ = a \cup \{a\} \in A$.

2.100   Theorem
There is a least (with respect to $\subseteq$) successor set.

2.101   Definition
The least (with respect to $\subseteq$) successor set is denoted by ω; the elements of ω are called **natural numbers** and are labeled as follows: $0 = \varnothing$, $1 = 0^+$, $2 = 1^+$, $3 = 2^+$, $\cdots$. For $a, b \in \omega$, we usually write $a \leq b$ instead of $a \subseteq b$.

2.102   **Theorem**
Let $S \subseteq \omega$ such that
  (a) $\emptyset \in S$, and
  (b) if $a \in S$, then $a^+ \in S$.
Then $S = \omega$.

This theorem *is* a triviality to prove, but it says in more familiar terms that finite induction "works" in ω.

2.103   **Theorem**
ω is an ordinal number.

2.104   **Proposition**
Let A and B be ordinal numbers. The set $(A \times \{0\}) \cup (B \times \{1\})$ is well-ordered by taking $x \leq y$ iff any one of the following holds
  (a) $x = (a, 0)$ and $y = (b, 1)$, or
  (b) $x = (a, 0)$ and $y = (a', 0)$ and $a \subseteq a'$, or
  (c) $x = (b, 1)$ and $y = (b', 1)$ and $b \subseteq b'$.

2.105   **Definition**
If A and B are ordinal numbers, then the **ordinal sum** $A + B$ is the ordinal number isomorphic to the well-ordered set of 2.104.

2.106   **Proposition**
Let A, B, and C be ordinal numbers.   Then
  (a) $A + 0 = 0 + A = A$.
  (b) $A + 1 = A^+$.
  (c) $A + (B + C) = (A + B) + C$.
  (d) $1 + \omega = \omega \neq \omega + 1$.

2.107   **Proposition**
  (a) If A and B are ordinal numbers, then $A \subseteq B$ iff there exists a unique ordinal number C such that $B = A + C$.
  (b) $\omega \subseteq \omega + 1$, but there is no ordinal number C for which $\omega + 1 = C + \omega$.

2.108   **Proposition**
Let A and B be ordinal numbers.   The set $A \times B$ is well-ordered by taking $(a, b) \leq (a', b')$ iff $b \subsetneq b'$ or $(b = b'$ and $a \subseteq a')$.

2.109   **Definition**
If A and B are ordinal numbers, then the **ordinal product** AB is the ordinal number isomorphic to the well-ordered set of 2.108.

2.110   ***Proposition***
Let A, B, and C be ordinal numbers.   Then
(a) $A0 = 0A = 0$.
(b) $A1 = 1A = A$.
(c) $A(BC) = (AB)C$.
(d) $2\omega = \omega \neq \omega + \omega = \omega2$.

2.111   ***Definition***
A **limit ordinal** is a nonzero ordinal number with no immediate predecessor.

2.112   ***Example***
ω, $\omega + \omega$, and $\omega\omega$ are limit ordinals.

2.113   ***Proposition***
If A is any ordinal number, then A can be written in the form $A = L + n$ where L is a limit ordinal or is zero and n is a natural number.

2.114   ***Proposition***
An ordinal number $A \neq 0$ is a limit ordinal iff A is the supremum of a set B of ordinal numbers that does not contain A.

2.115   ***Definition***
Two sets A and B are said to be **equipotent** (written $A \sim B$) iff there exists a bijection $F : A \to B$.

2.116   ***Example***
ω, $\omega + 1$, and $\omega + \omega$ are equipotent.   Later we will see (cf. 2.141) that each of these is equipotent to $\omega\omega$.

2.117   ***Proposition***
The relation $\sim$ is an equivalence relation on the class $\mathfrak{U}$ of all sets.

2.118   ***Proposition***
Let X be a set.   Then

$$\{A \mid A \text{ is an ordinal number and } A \sim X\}$$

is a nonempty set.

2.119   ***Definition***
If X is a set, then the **cardinal number of** X (written card X or $|X|$) is the least element of the set

$$\{A \mid A \text{ is an ordinal number and } A \sim X\}.$$

An ordinal number B is a **cardinal number** iff there is a set X such that $B = \operatorname{card} X$. If A and B are cardinal numbers, we usually write $A \leq B$ for $A \subseteq B$.

2.120   Examples

(a) $\operatorname{card}(\{\{\{\varnothing\}\}, \{\varnothing\}\}) = 2$.

(b) $\operatorname{card}(\omega + \omega) = \operatorname{card}(\omega + 1) = \operatorname{card} \omega = \omega$.

2.121   Theorem

The class $\mathfrak{C}$ of all cardinal numbers is a system of representatives for $\sim$ on $\mathfrak{U}$.

2.122   Definition

Let A and B be cardinal numbers. Then

(a) The **cardinal sum** of A and B is $A + B = \operatorname{card}(A \uplus B)$.

(b) The **cardinal product** of A and B is $A \cdot B = \operatorname{card}(A \times B)$.

(c) The **cardinal exponentiation** A to the power B is

$$A^B = \operatorname{card}(\{F \mid F : B \to A\}).$$

Note that the cardinal sum and product of two cardinal numbers are not necessarily the same as their ordinal sum and product. From now on it must be clear from the context or be made explicit which operations are intended.

2.123   Examples

Using the above definition, one can verify that

(a) $2^3 = 3^+ + 3^+ = 2 \cdot 3^+$.

(b) For any cardinal number A, $A^2 = A \cdot A$.

2.124   Proposition

If A and B are cardinal numbers, then $A \leq B$ iff A is equipotent with a subset of B.

2.125   Proposition

(a) If R is an equivalence relation on a set X, then $|X/R| \leq |X|$.

(b) If $f : A \to B$ is a surjection, then $|B| \leq |A|$.

(c) $|A \cup B| \leq |A| + |B|$, for all sets A and B.

2.126   Proposition

If $n \in \omega$, then $n = \operatorname{card} n = \operatorname{ord} n$, and on ω, cardinal and ordinal arithmetic agree.

2.127   Proposition

Let A, B, and C be cardinal numbers. Then, with respect to cardinal arithmetic,

(a) $A + B = B + A$ and $A \cdot B = B \cdot A$.

(b) $A + (B + C) = (A + B) + C$ and $A \cdot (B \cdot C) = (A \cdot B) \cdot C$.

(c) $A \cdot (B + C) = (A \cdot B) + (A \cdot C)$.

(d) $A^{(B + C)} = (A^B) \cdot (A^C)$.

(e) $A^{(B \cdot C)} = (A^B)^C$.

(f) $(A \cdot B)^C = (A^C) \cdot (B^C)$.

2.128   Proposition
If A, B, and C are cardinal numbers with $A \leq B$, then

(a) $A \cdot C \leq B \cdot C$.

(b) $A + C \leq B + C$.

(c) $C^A \leq C^B$.

(d) $A^C \leq B^C$.

2.129   Proposition
Let a, b, and c be natural numbers.   Then

(a) $a + b = b + a \in \omega$.

(b) $ab = ba \in \omega$.

(c) $a < b$ iff $a + c < b + c$.

(d) If $c \neq 0$, then $a < b$ iff $ac < bc$.

(e) there exists a unique $d \in \omega$ such that either $a = d + d$ or $a = (d + d) + 1$.

2.130   Theorem (Cantor-Bernstein)
If A and B are sets and if A is equipotent with a subset of B and B is equipotent with a subset of A, then A and B are equipotent.

2.131   Proposition
Let A be a set.   Then

(a) $|\mathscr{P}(A)| = 2^{|A|}$.

(b) $|A| \leq |\mathscr{P}(A)|$.

(c) $|A| \neq |\mathscr{P}(A)|$.

2.132   Definition
(a) If X is a set for which card $X \in \omega$, then X is said to be **finite**; otherwise, X is **infinite**.

(b) The smallest infinite cardinal number is written $\aleph_0$ ("aleph-zero"). A set X is **countable** if card $X \leq \aleph_0$; otherwise X is **uncountable**.

2.133   Example *(cf. 2.61)*
Card $\mathbf{N}$ = card $\mathbf{Q} = \aleph_0$, and card $\mathbf{R} = 2^{\aleph_0}$.

2.134   Proposition
(a) $\aleph_0 = \omega$.

(b) Every infinite set has a subset of cardinality $\aleph_0$.

(c) A set is infinite iff it is equipotent with a proper subset of itself.

2.135   **Examples**
(a) $\aleph_0 + \aleph_0 = \aleph_0$ while $\omega + \omega \neq \omega$.
(b) For each $i \in \omega$, let A_i be an ordinal number. Then it cannot be the case that $A_0 \supsetneqq A_1 \supsetneqq A_2 \supsetneqq \cdots \supsetneqq A_n \supsetneqq A_{n+1} \supsetneqq \cdots$.

2.136   **Proposition**
Each infinite cardinal number is a limit ordinal, but not conversely.

2.137   **Proposition**
$\aleph_0{}^2 = \aleph_0 = \aleph_0 \cdot \aleph_0$.

2.138   **Corollary**
A countable union of countable sets is countable.

2.139   **Lemma**
(a) If A is an infinite cardinal number for which $A = A \cdot A$, then

$$A = 2 \cdot A = 3 \cdot A = A \cdot A.$$

(b) If X is an infinite set and $A \subseteq X$ such that $|A| = |A| \cdot |A|$ and $|A| < |X|$, then $|A| < |X - A|$.

2.140   **Theorem**
If A is any infinite cardinal number, then $A = A^2$.

2.141   **Theorem**
If A and B are cardinal numbers with A infinite and $B \neq 0$, then

$$A + B = A \cdot B = \text{the larger of } A \text{ and } B.$$

2.142   **Corollary**
If X is an infinite set and $\mathrm{Fin}(X)$ is the set of all finite subsets of X, then card $X =$ card $\mathrm{Fin}(X)$.

2.143   **Corollary**
If X is an infinite set and $A \subseteq X$ such that card $A <$ card X, then card$(X - A) =$ card X.

2.144   **Corollary**
If A and B are cardinal numbers with B infinite, and if $2 \leq A \leq B$, then $A^B = 2^B$.

2.145   **Proposition**
For each cardinal number A, there is a cardinal number B such that $A < B$ and for no cardinal number C it is true that $A < C < B$.

2.146   Definition

If A is an infinite cardinal number and if B is the unique ordinal number isomorphic to the well-ordered set

$$\{C \mid C \text{ is an infinite cardinal number and } C < A\},$$

we write $A = \aleph_B$.

2.147   Proposition

(a) The definitions of $\aleph_0$ in 2.146 and 2.132 agree.

(b) If $\aleph_A$ is an infinite cardinal number, then the "next larger" cardinal number is $\aleph_{A+1}$.

(c) For every ordinal number A, there is a cardinal number $\aleph_A$.

2.148   Theorem

The class of all ordinal numbers is isomorphic to the class of all infinite cardinal numbers.

2.149   Theorem

Let Ω denote the first uncountable ordinal number (i.e., Ω is the set of all countable ordinal numbers). If A is any countable subset of Ω, then A has an upper bound in Ω.

2.150   Proposition

$\aleph_1 = \Omega$.

2.151

The following two statements are independent of our axioms (Axioms 1 through 9)—this was proved by Gödel (1940, consistency) and Cohen (1963, consistency of negation). It is not felt, as was the case with the axiom of choice, that there is everything to gain and nothing to lose by assuming them. The analogy here is often made with Euclid's parallel postulate, in that interesting mathematics may come from assuming some form of the negation of the generalized continuum hypothesis. Any theorem whose proof depends upon the continuum hypothesis should explicitly say so in its statement.

Generalized Continuum Hypothesis

For each infinite cardinal number $\aleph_A$, $2^{\aleph_A} = \aleph_{A+1}$.

Continuum Hypothesis

$2^{\aleph_0} = \aleph_1$.

Intuitively, the continuum hypothesis says that there is no subset of the real line (which classically is called the **continuum**) with more elements than the integers and fewer elements than the whole line.

CHAPTER 3

THE REAL LINE

In this chapter we present a development of the set **R** of real numbers based only on the set theory of Chapter 2. Although our treatment of **R** is far from exhaustive, enough of its geometric and algebraic properties are exhibited to show that its development is worth the work. We also briefly consider convergence and summability of sequences and use these concepts to define decimal expansions. All of this is needed (at some level of sophistication) in Chapter 4.

The beauty and significance of this development lie in the fact that it can be carried out, with relative ease, on the basis of the axioms of set theory. Many of the *details* of the proofs to be encountered in this chapter are long, uninteresting, and unenlightening. Thus we suggest that in most cases proofs only be sketched out to the point that the calculations begin to overwhelm. We do not feel, however, that the chapter should be omitted. It is an integral part of the development, and specific facts about **R** are needed in an essential way in topology.

3.1

For a, $b \in \omega$, the equation $a + x = b$ can be solved if and only if $b \geq a$. Since we want to be able to solve this equation for any values of a and b, we shall extend our horizons by introducing "artificial" solutions whenever necessary. After several definitions and some calculation, this yields the familiar set **Z** of all integers.

3.2 **Definition**

(a) Let S be the equivalence relation on $\omega \times \omega$ defined by

$$S = \{((n, m), (p, q)) \mid n + q = p + m \text{ in } \omega\}.$$

The quotient set $(\omega \times \omega)/S$ is denoted by **Z** and we call the members o **Z integers**.

To simplify notation, we shall write $\langle n - m \rangle$ instead of $S[\{(n, m)\}]$.

(b) Let $\leq$ be the relation on **Z** given by

$$\langle n - m \rangle \leq \langle p - q \rangle \quad \text{iff} \quad n + q \leq p + m \text{ in } \omega.$$

The element $\langle n - m \rangle$ will turn out to be a solution to the equation $m + x = n$, i.e., to represent the "intuitive integer" $n - m$. You should keep clearly in mind, however, that $\langle n - m \rangle$ is a formal symbol—an abbreviation—and that "$-$" here is not considered as an operation on the set of natural numbers.

It is important to prove that the above "definition" of $\leq$ really does define a relation. Since it *can* happen that $\langle n - m \rangle = \langle n' - m' \rangle$ when $n \neq n'$ and $m \neq m'$, we must be certain that in this case $\langle n - m \rangle \leq \langle p - q \rangle$ iff $\langle n' - m' \rangle \leq \langle p - q \rangle$, for any integer $\langle p - q \rangle$. The following is a proof that $\leq$ is "well-defined".

Suppose that $\langle n - m \rangle = \langle n' - m' \rangle$. By the definition of the equivalence relation S, we have $n + m' = n' + m$. Now, $\langle n - m \rangle \leq \langle p - q \rangle$ iff $n + q \leq p + m$ and this is equivalent to $n + q + (m' + n') \leq p + m + (m' + n')$. Using commutativity and associativity in ω (2.106 and 2.129), this is equivalent to $(n + m') + (n' + q) \leq (n' + m) + (p + m')$ which, in view of 2.129, is the same as $n' + q \leq p + m'$, or $\langle n' - m' \rangle \leq \langle p - q \rangle$. In a similar way, the definition is independent of the representative chosen for $\langle p - q \rangle$.

3.3   Theorem
$(\mathbf{Z}, \leq)$ is a totally ordered set, and the function $n \mapsto \langle n - 0 \rangle$ from ω to $\mathbf{Z}$ is injective and order-preserving, i.e., is an order isomorphism onto its image $\{z \in \mathbf{Z} \mid z \geq \langle 0 - 0 \rangle\}$.

3.4   Definition
We write $\mathbf{N}$ for the set $\{z \in \mathbf{Z} \mid z > \langle 0 - 0 \rangle\}$ and call the elements of $\mathbf{N}$ **positive integers**. If $z < \langle 0 - 0 \rangle$, we say that z is a **negative integer**.

3.5   Definition
(a) A **group** is an ordered pair $(G, +)$, where $+$ is a binary operation on G [see 2.39(d)] satisfying
   (1) $+$ is **associative**: $(a + b) + c = a + (b + c)$ for all $a, b, c \in G$;
   (2) $+$ has an **identity**, e: there exists an $e \in G$ such that $a + e = e + a = a$ for all $a \in G$; and
   (3) each element of G has an **inverse**: for each $a \in G$, there exists an element of G (denoted by) $-a$ such that $a + (-a) = (-a) + a = e$.
The element $a + b$ is called the **sum** of a and b.

(b) A group $(G, +)$ is called **abelian** (or **commutative**) provided that $a + b = b + a$ for all $a, b \in G$.

3.6   Theorem
Let $+$ be the binary operation on $\mathbf{Z}$ given by

$$\langle n - m \rangle + \langle p - q \rangle = \langle (n + p) - (m + q) \rangle.$$

Then $(\mathbf{Z}, +)$ is an abelian group with identity $\langle 0 - 0 \rangle$, where the inverse of an element $\langle n - m \rangle$ is $\langle m - n \rangle$. Moreover, the function $n \mapsto \langle n - 0 \rangle$ from ω to $\mathbf{Z}$ preserves sums [i.e., $i(n + m) = i(n) + i(m)$, where the first "$+$" is ordinal addition and the second "$+$" is that defined above].

3.7 Corollary

For any a, $b \in \mathbf{Z}$, the equation $a + x = b$ can be solved in $\mathbf{Z}$.

3.8 Definition

(a) A **ring** is an ordered triple $(A, +, \cdot)$ where $+$ and $\cdot$ are binary operations on A satisfying

(1) $(A, +)$ is an abelian group (whose identity is usually denoted by 0),

(2) $\cdot$ is associative, and

(3) $\cdot$ **distributes over** $+$: for all a, b, $c \in A$,

$$a \cdot (b + c) = (a \cdot b) + (a \cdot c),$$
$$(b + c) \cdot a = (b \cdot a) + (c \cdot a).$$

The element $a \cdot b$ is called the **product** of a and b.

(b) A ring $(A, +, \cdot)$ is said to be **commutative** provided that $a \cdot b = b \cdot a$ for all a, $b \in A$.

(c) A ring $(A, +, \cdot)$ is called a ring **with identity** provided that there exists an element $u \in A$ (called an **identity**) such that $u \cdot a = a \cdot u = a$ for all $a \in A$.

We shall frequently omit superfluous parentheses. Associativity of $+$ and $\cdot$ permit us to write $a + b + c$ or $a \cdot b \cdot c$ without ambiguity. We also make the convention that products take precedence over sums, so that for instance $a + b \cdot c$ means $a + (b \cdot c)$ and *not* $(a + b) \cdot c$.

3.9 Theorem

Let $\cdot$ be the binary operation on $\mathbf{Z}$ given by

$$\langle n - m \rangle \cdot \langle p - q \rangle = \langle (np + mq) - (nq + mp) \rangle.$$

Then $(\mathbf{Z}, +, \cdot)$ is a commutative ring with identity $\langle 1 - 0 \rangle$. Moreover, the function $n \mapsto \langle n - 0 \rangle$ from ω to $\mathbf{Z}$ preserves products.

Since the function $n \mapsto \langle n - 0 \rangle$ from ω to $\mathbf{Z}$ is injective and preserves all the structures of interest (i.e., order, sums, and products), we shall no longer distinguish between ω and its copy in $\mathbf{Z}$. Henceforth we will write n for $\langle n - 0 \rangle$ and $-n$ for $\langle 0 - n \rangle$.

3.10 Proposition

Let a, b, $c \in \mathbf{Z}$.

(a) If $a \leq b$, then $a + c \leq b + c$.

(b) If $a \geq 0$ and $b \geq 0$, then $a \cdot b \geq 0$.

(c) There exists an $n \in \omega$ such that either $a = n$ or $a = -n$.

(d) If $a \cdot b = 0$, then either $a = 0$ or $b = 0$.

3.11

We have now extended ω to make a new set **Z** in which we can subtract, and we have in a natural way extended the notions of (ordinal) sum and product. It is important to note, however, that **Z** is not well-ordered.

Z still has defects to be patched up (or viewed positively, there are new and more interesting numbers to consider). For example, in **Z**, one cannot always solve the equation $a \cdot x = b$ for x. As before, we just introduce solutions to such equations and then see what happens. One point requires comment. If $a = 0$ and $b \neq 0$, the above equation cannot have any solution in a ring, since $0 \cdot x = 0$ in every ring. On the other hand, if $a = b = 0$, then any x is a solution.

3.12   Definition

(a) Let T be the equivalence relation on $\mathbf{Z} \times \mathbf{N}$ defined by

$$T = \{((a, b), (c, d)) \mid a \cdot d = c \cdot b \text{ in } \mathbf{Z}\}.$$

The quotient set $(\mathbf{Z} \times \mathbf{N})/T$ is denoted by **Q** and we call the members of **Q rational numbers**.

To simplify notation, we will write $\langle a/b \rangle$ instead of $T[\{(a, b)\}]$.

(b) Let $\leq$ be the relation on **Q** given by

$$\langle a/b \rangle \leq \langle c/d \rangle \quad \text{iff} \quad a \cdot d \leq c \cdot b \text{ in } \mathbf{Z}.$$

The element $\langle b/a \rangle$ will turn out to be the solution to the equation $a \cdot x = b$, for $a > 0$, i.e., to represent the "intuitive rational number" b/a. However, as in the earlier situation with integers, $/$ should not be considered as an operation on integers. The symbol $\langle b/a \rangle$ must be taken as a whole, as the equivalence class to which (b, a) belongs.

3.13   Theorem

$(\mathbf{Q}, \leq)$ is a totally ordered set and the function $z \mapsto \langle z/1 \rangle$ from **Z** to **Q** is an order isomorphism onto its image.

3.14   Theorem

Let $+$ be the binary operation on **Q** given by

$$\langle a/b \rangle + \langle c/d \rangle = \langle (a \cdot d + c \cdot b)/b \cdot d \rangle.$$

Then $(\mathbf{Q}, +)$ is an abelian group with identity $\langle 0/1 \rangle$, where the inverse of an element $\langle a/b \rangle$ is $\langle (-a)/b \rangle$. Moreover, the function $z \mapsto \langle z/1 \rangle$ from **Z** to **Q** preserves sums.

3.15   Theorem

Let $\cdot$ be the binary operation on **Q** given by $\langle a/b \rangle \cdot \langle c/d \rangle = \langle a \cdot c \mid b \cdot d \rangle$. Then $(\mathbf{Q}, +, \cdot)$ is a commutative ring with identity $\langle 1/1 \rangle$. Moreover, the function $z \mapsto \langle z/1 \rangle$ from **Z** to **Q** preserves products.

3.16   **Definition**
A **field** is a commutative ring $(F, +, \cdot)$ with identity $1 \neq 0$ such that if $0 \neq r \in F$, then the equation $r \cdot x = 1$ has a solution. This solution is denoted by r^{-1} and is called the (**multiplicative**) **inverse** of r.

In a field, any equation $a \cdot x = b$ $(a \neq 0)$ can be solved: $x = a^{-1} \cdot b$ is a solution.

3.17   **Theorem**
$(\mathbf{Q}, +, \cdot)$ is a field in which the inverse of $\langle a/b \rangle$ $(a \neq 0)$ is $\langle b/a \rangle$ if $a > 0$, and is $\langle -b/-a \rangle$ if $a < 0$.

3.18   **Corollary**
If $a, b \in \mathbf{Q}$ and $a \neq 0$, then the equation $a \cdot x = b$ can be solved in $\mathbf{Q}$.

Since the function $z \mapsto \langle z/1 \rangle$ from $\mathbf{Z}$ to $\mathbf{Q}$ is injective and preserves order, sums, and products, we will no longer distinguish between $\mathbf{Z}$ and its copy in $\mathbf{Q}$. Henceforth we shall write z for $\langle z/1 \rangle$ and a/b or $\frac{a}{b}$ for $\langle a/b \rangle$.

We have thus extended $\mathbf{Z}$ to a structure in which "division" is possible.

3.19   **Proposition**
Let $a, b, c \in \mathbf{Q}$.
    (a) If $a \leq b$, then $a + c \leq b + c$.
    (b) If $a \geq 0$ and $b \geq 0$, then $a \cdot b \geq 0$.

3.20   **Theorem**
    (a) $\mathbf{Q}$ is **archimedean** (i.e., if $a, b \in \mathbf{Q}$ and if $n \cdot a \leq b$ for all $n \in \mathbf{N}$, then $a \leq 0$).
    (b) No proper subset of $\mathbf{Q}$ (whose operations are the restrictions of those on $\mathbf{Q}$) is a field.
    (c) if $a > b$ in $\mathbf{Q}$, then there exists some $q \in \mathbf{Q}$ such that $a > q > b$.

3.21

A **totally ordered field** is a 4-tuple $(F, +, \cdot, \leq)$ where $(F, +, \cdot)$ is a field, $(F, \leq)$ is a totally ordered set, and the conditions
    (a) if $a \leq b$, then $a + c \leq b + c$
    (b) if $a \geq 0$ and $b \geq 0$, then $a \cdot b \geq 0$
of Proposition 3.19 are satisfied.

If $(F, +, \cdot, \leq)$ is a totally ordered field no proper subset of which is a field, then there is an order isomorphism from F onto $\mathbf{Q}$ which preserves the ring operations, so that in this case F and $\mathbf{Q}$ are essentially the same. Hence $\mathbf{Q}$ is characterized as the smallest totally ordered field.

3.22

Although $\mathbf{Q}$ has some properties that are much nicer than those of $\mathbf{Z}$, there are still some defects.   For example, in $\mathbf{Q}$ there is no solution to the equation $x \cdot x = 2$.   Also, $\mathbf{Q}$ has many "gaps" in its order; specifically, between any two different rational numbers there is a subset of $\mathbf{Q}$ that has no supremum in $\mathbf{Q}$.   We shall see that by filling the gaps in $\mathbf{Q}$, we automatically take care of the first difficulty.

If we denote by $\mathbf{R}$ the totally ordered field that we get by filling in all gaps in $\mathbf{Q}$, then $\mathbf{R}$ will be the largest archimedean totally ordered field. No further extension is possible without sacrificing something significant.

As before, we just manufacture what we need.

3.23   *Definition*

(a) A **cut** (or **Dedekind cut**) in $\mathbf{Q}$ is a nonempty proper ideal of $(\mathbf{Q}, \leq)$ that has no greatest element.

(b) We denote by $\mathbf{R}$ the set of all cuts in $\mathbf{Q}$; elements of $\mathbf{R}$ are called **real numbers**, and $\mathbf{R}$ is called the **real line**.

3.24   *Theorem*

(a) $(\mathbf{R}, \subseteq)$ is a totally ordered set.

(b) Every nonempty subset of $\mathbf{R}$ that has an upper bound has a supremum and every nonempty subset of $\mathbf{R}$ that has a lower bound has an infimum [i.e., $(\mathbf{R}, \subseteq)$ is **conditionally complete**].

It is conventional to write $\leq$ rather than $\subseteq$ for the total order on $\mathbf{R}$.

3.25   *Theorem*

(a) The function $i : \mathbf{Q} \to \mathbf{R}$ given by $i(q) = \{x \in \mathbf{Q} \,|\, x < q\}$ is an order isomorphism of $\mathbf{Q}$ onto its image.

(b) If S is any subset of $\mathbf{Q}$ that has a supremum s, then $i(s) = \sup i[S]$; similarly for infimum.

3.26   *Theorem*

If $r, r' \in \mathbf{R}$ with $r > r'$, then there exists some $q \in \mathbf{Q}$ such that $r > i(q) > r'$.

3.27   *Definition*

Let $r \in \mathbf{R}$.   Denote by $m(r)$ the cut obtained by removing the greatest element (if there is one) from the set

$$\{x \in \mathbf{Q} \,|\, x \leq -q \text{ for all } q \in r\}.$$

The **absolute value** of r is the cut $|r| = r \cup m(r)$.

3.28   *Proposition*

For $r \in \mathbf{R}$, $r \geq i(0)$ iff $r = |r|$.

3.29   ***Theorem***
Let $+$ be the binary operation on $\mathbf{R}$ given by

$$r + s = \{x + y \mid x \in r \text{ and } y \in s\}.$$

Then $(\mathbf{R}, +)$ is an abelian group with identity $i(0)$, where the inverse of an element r is $m(r)$. Moreover, the function $i : \mathbf{Q} \rightarrow \mathbf{R}$ given by $i(q) = \{x \in \mathbf{Q} \mid x < q\}$ preserves sums.

We denote the additive identity of $\mathbf{R}$ by 0 and the additive inverse of $r \in \mathbf{R}$ [i.e., $m(r)$] by $-r$.

3.30   ***Definition***
Let $r, s \in \mathbf{R}$.
   (a) If $r \geq 0$ and $s \geq 0$, then the **product** of r and s is the cut

$$r \cdot s = \{x \in \mathbf{Q} \mid \text{ for some } 0 \leq p \in r \text{ and } 0 \leq q \in s, x \leq p \cdot q\} \cup 0.$$

(Remember that 0 is the cut $\{q \in \mathbf{Q} \mid q < 0\}$.)
   (b) In general, the product of r and s is given by

$$r \cdot s = \begin{cases} |r| \cdot |s| & \text{if } r \geq 0 \text{ and } s \geq 0, \text{ or if } r \leq 0 \text{ and } s \leq 0, \\ -(|r| \cdot |s|) & \text{otherwise.} \end{cases}$$

3.31   ***Theorem***
   (a) $(\mathbf{R}, +, \cdot)$ is a field with identity $i(1)$.
   (b) The function $i : \mathbf{Q} \rightarrow \mathbf{R}$ given by $i(q) = \{x \in \mathbf{Q} \mid x < q\}$ preserves products.

Since i preserves order, sums, and products, we shall no longer distinguish between $\mathbf{Q}$ and its copy in $\mathbf{R}$.

3.32   ***Example***
Let $0 \leq r \in \mathbf{R}$.   Then

$$\sqrt{r} = \{q \in \mathbf{Q} \mid q < 0 \text{ or } q \cdot q < r\}$$

is a cut whose **square** (i.e., product with itself) is r.   Hence $\mathbf{R}$ contains **square roots** of its positive elements.

3.33   ***Theorem***
$\mathbf{R}$ is an archimedean totally ordered field that is conditionally complete.

3.34   ***Remark***
Up to an isomorphism of all structures, $\mathbf{R}$ is the only conditionally complete (archimedean) totally ordered field.

This completes the development of **R**. We now consider several properties of **R** which will be needed in Chapter 4.

3.35   Definition

Let X be any set. A **sequence in** X is a function from either ω or **N** to X (recall that ω contains 0 and **N** does not). Since a sequence is a family, we usually write a_n instead of $a(n)$, $(a_n)_{n \in \omega}$ instead of $a : \omega \to X$, and $(a_n)_{n \in \mathbf{N}}$ instead of $a : \mathbf{N} \to X$. When no confusion is likely, we drop the subscripts $n \in \omega$ and $n \in \mathbf{N}$.

3.36   Examples

(a) The sequence $(1/n)_{n \in \mathbf{N}}$ is the function $f : \mathbf{N} \to \mathbf{R}$ given by $f(n) = 1/n$.

(b) The set of all rational numbers is the image of some sequence since, by 2.138, **Q** is countable.

(c) The sequence $(a)_{n \in \omega}$ in a set X is the function $f : \omega \to X$ whose value is always a. Such a sequence is called a **constant sequence**.

3.37   Definition

A sequence (a_n) in **R** is said to **converge to** $r \in \mathbf{R}$ (and we say that r is the **limit** of the sequence) provided that for each positive real number ε, there exists some $n_0 \in \omega$ (n_0 depending on ε) such that $|a_m - r| < \varepsilon$ whenever $m \geq n_0$. In this case we write $r = \lim_{n \to \infty} a_n$ or $r = \lim(a_n)$.

A sequence in **R** is said to **converge** provided that there exists an $r \in \mathbf{R}$ such that the sequence converges to r.

3.38   Examples

(a) Since **R** is archimedean, the sequence $(1/n)_{n \in \mathbf{N}}$ converges to 0.

(b) No sequence whose image is all of **Q** can converge.

(c) Any constant sequence converges to its constant value.

3.39   Definition

(a) A sequence (a_n) in **R** is said to be **monotone increasing** provided that $a_n \geq a_m$ whenever $n > m$; (a_n) is termed **monotone decreasing** if $a_n \leq a_m$ whenever $n > m$.

(b) A sequence in **R** is **bounded** provided that its image has an upper bound and a lower bound in **R**.

3.40   Theorem

Every bounded monotone increasing (or monotone decreasing) sequence in **R** converges.

3.41   Definition

A sequence (a_n) in **R** is said to be a **Cauchy sequence** provided that for each real number $\varepsilon > 0$ there exists an $n \in \omega$ such that $|a_m - a_{m'}| < \varepsilon$ whenever $m, m' > n$.

3.42   Proposition
Every convergent sequence in **R** is Cauchy.

3.43   Theorem
Every Cauchy sequence in **R** converges.

3.44   Definition
Let $(t_n)_{n \in \omega}$ be a sequence in **R**. The associated sequence of **partial sums** is the sequence

$$\left(\sum_{i=0}^{n} t_i \right)_{n \in \omega} \qquad \text{where } \sum_{i=0}^{n} t_i = t_0 + t_1 + \cdots + t_n.$$

If the sequence of partial sums converges, then we say that (t_n) is **summable**, its limit is denoted by

$$\sum_{i=0}^{\infty} t_i \qquad \text{or} \qquad \sum_{i \in \omega} t_i,$$

and its limit is called the **sum** of the sequence $(t_n)_{n \in \omega}$.

Similar statements hold for sequences indexed by **N** instead of ω.

3.45   Example
For any $n \in \mathbf{N}$ and $r \in \mathbf{R}$, let r^n denote the real number $r \cdot r \cdot \ldots \cdot r$ (n times); or, phrased inductively, $r^0 = 1$, $r^1 = r$, and if r^k has been defined, $r^{k+1} = r \cdot r^k$. Then $\sum_{i=0}^{\infty} 1/2^n = 2$.

3.46   Definition
(a) Let r be a nonnegative real number. A sequence $(d_n)_{n \in \omega}$ is said to be a **decimal expansion for** r provided that the following conditions are satisfied:

(1) d_i is a natural number for all $i \in \omega$.
(2) $0 \leq d_i < 10$ for all $i \in \mathbf{N}$.
(3) The sequence $(d_i/10^i)_{i \in \omega}$ is summable.
(4) $r = \sum_{i=0}^{\infty} d_i/10^i$.

If $(d_n)_{n \in \omega}$ is a decimal expansion for r, then we write

$$r = d_0.d_1 d_2 d_3 \cdots.$$

(b) If $r \in \mathbf{R}$ is negative and if $(d_n)_{n \in \omega}$ is a decimal expansion for $|r|$, then we write

$$r = -d_0.d_1 d_2 d_3 \cdots.$$

3.47   Examples
(a) $\frac{1}{8} = 0.12500 \cdots$. In the future, superfluous zeros will be suppressed.

(b) $\frac{1}{3} = .3333 \cdots$.

(c) $1 = .99999999 \cdots$.

3.48   Theorem

Let S be the set of all sequences $(d_n)_{n \in \omega}$ in $\mathbf{R}$ such that

(a) each d_n is a natural number,

(b) $0 \leq d_n < 10$ for all $n \in \mathbf{N}$, and

(c) for each $n \in \mathbf{N}$, there exists an $m \geq n$ such that $d_m \neq 9$.

Then

(1) for every $(d_n)_{n \in \omega}$ in S, the sequence $(d_i/10^i)_{i \in \omega}$ is summable, and

(2) the function $D : S \rightarrow \mathbf{R}$ defined by

$$D((d_n)_{n \in \omega}) = \sum_{i=0}^{\infty} d_i/10^i$$

is one-to-one and onto the set of nonnegative real numbers.

Thus we have that every nonnegative real number has a unique decimal expansion that is not ultimately constant 9's.

3.49

By replacing the number 10 above (3.46 and 3.48) by any other integer $b > 1$ (and replacing 9 by $b - 1$ in the theorem) we obtain the result: Every nonnegative real number has a unique expansion base b that is not ultimately constant $(b - 1)$'s.

3.50   Examples

(a) Even though $\mathbf{N}$, $\mathbf{Z}$, and $\mathbf{Q}$ are countable, $\mathbf{R}$ is uncountable. Indeed, card $\mathbf{R} = 2^{\aleph_0}$.

(b) An interesting question: Does there exist an uncountable set of subsets of $\mathbf{N}$ that is totally ordered by inclusion?

CHAPTER 4

TOPOLOGY

Except in some of the examples, we consider topological spaces as purely abstract structures. However, the *motivation* for many topological notions and theorems lies in analysis. An examination of the structure necessary to reflect the geometry of **R** will help us to obtain the proper perspective. We shall thus return to elementary calculus for a moment to review the important notion of continuity, the generalization of which is the keystone of topology.

A reasonable series of approximations to a rigorous definition of the notion of a continuous real-valued function of a real variable might go as follows:

(a) A function is continuous iff its graph is a continuous curve.

(b) A function is continuous iff its graph can be drawn without lifting the pencil.

(c) A function f is continuous iff for every x in its domain, $f(y)$ is close to $f(x)$ whenever y is close to x.

(d) A function f is continuous iff for every x in the domain of f $|f(x) - f(y)|$ can be made very small by choosing $|x - y|$ small enough.

(e) A function f is continuous iff it is continuous at each element of its domain; and it is continuous at the point x iff for every $\varepsilon > 0$, there exists a $\delta > 0$ such that $|f(y) - f(x)| < \varepsilon$ whenever y is in the domain of f with $|y - x| < \delta$.

With respect to the present-day idea of rigor, only (e) is a precise formal definition, but the other statements give a feeling for what is going on.

A careful examination of these statements shows that in each case, the central idea is "closeness". Topology encompasses all attempts to abstract the notion of "closeness" so that it can be used in places other than the real line, the plane, and higher euclidean spaces.

Historically, the first successful abstraction (these things are not God-given; an abstraction is considered successful if it gives the theorems that you want it to) was the idea of a metric space. A metric space is a set together with a way of measuring "distance" between any two points, subject to the following restrictions:

If X is the set, and $d(x, y)$ denotes the "distance between" x and y, we require (only) that

(a) $d(x, y) = d(y, x) \geq 0$,

(b) $d(x, y) = 0$ iff $x = y$, and

(c) $d(x, y) + d(y, z) \geq d(x, z)$.

The function $d : X \times X \to \mathbf{R}$ is called a metric on X. To see the motivation for this definition (other than that it does give the "right" theorems), observe that

$$(x, y) \mapsto |x - y|$$

is a metric on $\mathbf{R}$, and that

$$((x_1, y_1), (x_2, y_2)) \mapsto \sqrt{(x_1 - x_2)^2 + (y_1 - y_2)^2}$$

is a metric on $\mathbf{R}^2$.

The second successful abstraction, and the one principally of interest to us here, is somewhat less familiar. If we do not want to measure closeness numerically, as in the case of metric spaces, just how can we do it? For help with this, let's go back to the first example and change the way of saying part (e) by introducing some convenient notation. Let

$$S(x, \varepsilon) = \{y \in \mathbf{R} \mid |x - y| < \varepsilon\}$$

and let

$$N_x = \{S(x, \varepsilon) \mid \varepsilon > 0\}.$$

We can rephrase (e) as follows: f is continuous iff for every x in the domain of f and $S \in N_{f(x)}$, there exists some $T \in N_x$ such that $f[T] \subseteq S$.

The thing of main concern here is the set N_x of "ε-neighborhoods" of x. Loosely speaking, the set of neighborhoods of x tells "how close" to x other elements of $\mathbf{R}$ are.

One further step. We shall call a set U of real numbers open iff each point $x \in U$ has some neighborhood $S(x, \varepsilon)$ which is contained in U. It can be verified that the set of all open subsets of $\mathbf{R}$ satisfies the following conditions:

(a) $\varnothing$ and $\mathbf{R}$ are open.

(b) If U and V are open, then $U \cap V$ is open.

(c) If $\mathscr{W}$ is a set of open sets, then $\bigcup \mathscr{W}$ is open.

We abstract the idea of "the set of all open subsets" to mean (a), (b), and (c), and this abstraction yields a rich and pleasing theory. This is a topology.

To conclude, let us return again to calculus, where we recall the following theorems:

Intermediate Value Theorem
If f is continuous on an interval containing a and b, and if $f(a) < c < f(b)$, then there exists an x between a and b such that $c = f(x)$.

Extreme Value Theorem
If f is continuous on a closed and bounded interval, then f assumes its supremum and infinum values on this interval.

Most often in calculus texts, the proofs of these important theorems (and several others) are omitted with the explanation that they are "beyond the scope of this book ". The scope of topology, however, is broad enough to make these theorems easy corollaries of general theorems which are themselves quite easy to prove (4.147 and 4.157).

4.1 **Definition**

A **topology** on a set X is a set τ of subsets of X for which

    (a) $\varnothing \in \tau$ and $X \in \tau$,

    (b) if $U, V \in \tau$, then $U \cap V \in \tau$, and

    (c) if $\mathscr{U} \subseteq \tau$, then $\bigcup \mathscr{U} \in \tau$.

A (**topological**) **space** is a pair (X, τ), where τ is a topology on X. If (X, τ) is a space, the elements of τ are called τ-**open sets** (or just **open sets**), and the elements of X are called **points** of the space. When no confusion is likely, the space (X, τ) will simply be denoted by X.

4.2 **Examples**

    (a) Let X be any set, and let $\tau = \mathscr{P}(X)$. Then τ is a topology on X (the **discrete topology**), and every subset of X is open.

    (b) Let X be any set, and let $\tau = \{\varnothing, X\}$. Then τ is a topology on X (the **indiscrete topology**).

    (c) Let $\mathbf{R}$ be the set of real numbers and let τ consist of all subsets of $\mathbf{R}$ that can be written as a union of **open intervals**, i.e., intervals of the form

$$(a, b) = \{x \mid a < x < b\} \qquad \text{for some } a, b \in \mathbf{R}.$$

Then τ is a topology on $\mathbf{R}$ (the **usual topology** on $\mathbf{R}$).

PROOF   Clearly $\varnothing = (1, 1) \in \tau$ and $\mathbf{R} = \bigcup\{(-n, n) \mid n \in \omega\} \in \tau$. Hence 4.1(a) is satisfied. If

$$U = \bigcup\{(a_\alpha, b_\alpha) \mid \alpha \in A\}$$
$$V = \bigcup\{(a_\beta, b_\beta) \mid \beta \in B\},$$

then

$$U \cap V = \bigcup\{(a_\alpha, b_\alpha) \cap (a_\beta, b_\beta) \mid \alpha \in A \text{ and } \beta \in B\}$$
$$= \bigcup\{(a_{\alpha, \beta}, b_{\alpha, \beta}) \mid \alpha \in A \text{ and } \beta \in B\},$$

where $a_{\alpha, \beta}$ is the larger of a_α and a_β, and $b_{\alpha, \beta}$ is the smaller of b_α and b_β. Hence 4.1(b) is satisfied. Clearly a union of unions of open intervals is a union of open intervals. Hence 4.1(c) is satisfied.

    (d) Let X be any infinite set and let τ consist of all subsets of X whose complement is finite (together with $\varnothing$). Then τ is a topology on X (the **cofinite topology**).

4.3 **Definition**

Let X be a set and let τ and σ be topologies on X. We say that τ is **finer** (or **stronger**) than σ (and that σ is **coarser** (or **weaker**) than τ) iff $\tau \supseteq \sigma$.

4.4   Proposition

The set of all topologies on a set is partially ordered by $\subseteq$.

4.5   Examples

(a) The discrete topology is the finest topology on any set and the indiscrete topology is the coarsest.

(b) On **R**, the topologies of 4.2 are ordered as follows: (discrete) $\supseteq$ (usual) $\supseteq$ (cofinite) $\supseteq$ (indiscrete).

(c) The set of topologies on any set with more than one element is not totally ordered by $\subseteq$.

4.6   Definition

Let X be a space. A **neighborhood** of a point $x \in X$ is any set H that contains an open set V containing x:

$$x \in V \subseteq H, \qquad V \text{ open.}$$

4.7   Proposition

Let X be a space. A subset U of X is open iff each point of U has a neighborhood contained in U.

4.8   Definition

(a) Let (X, τ) be a space, and let $\mathscr{B}$ be a set of subsets of X. Then $\mathscr{B}$ is said to be a **base for** τ iff

(1) $\mathscr{B} \subseteq \tau$, and

(2) every element of τ is the union of a subset of $\mathscr{B}$.

If $\mathscr{B}$ is a base for τ, then τ is called the **topology generated by** $\mathscr{B}$.

(b) Let X be a space and let $x \in X$. A set N_x of neighborhoods of x is called a **neighborhood base at** x iff every neighborhood of x contains a member of N_x.

4.9   Proposition

Let (X, τ) be a space and let $\mathscr{B} \subseteq \mathscr{P}(X)$. Then $\mathscr{B}$ is a base for τ iff

(a) $\mathscr{B} \subseteq \tau$, and

(b) for each $U \in \tau$ and $x \in U$, there exists some $B \in \mathscr{B}$ such that $x \in B \subseteq U$.

4.10   Proposition

Let X be a set and let $\mathscr{B} \subseteq \mathscr{P}(X)$. Then $\mathscr{B}$ is a base for *some* topology on X iff

(a) $X = \bigcup \mathscr{B}$, and

(b) whenever $U, V \in \mathscr{B}$ and $x \in U \cap V$, then there exists some $W \in \mathscr{B}$ such that $x \in W \subseteq U \cap V$.

4.11   Examples

(a) Every topology τ is a base for itself.

(b) Let X be a set and τ be the discrete topology on X. Each of the following is a base for τ:

    (1) $\{\{x\} \mid x \in X\}$.

    (2) $\{S \in \mathscr{P}(X) \mid \text{card } S \leq 3\}$.

    (3) $\mathscr{P}(X)$.

(c) The set of open intervals in **R** is a base for the usual topology on **R**.

(d) Let (X, d) be any **metric space** [i.e., X is a set and $d : X \times X \to \mathbf{R}$ such that

    (1) $d(x, y) = 0$ iff $x = y$,

    (2) $d(x, y) = d(y, x)$, and

    (3) $d(x, y) + d(y, z) \geq d(x, z)$].

Let $\mathscr{B}$ be the set of all **open balls** in (X, d) [i.e., sets of the form $\{x \in X \mid d(x, c) < r\}$, for some $c \in X$ and $0 < r \in \mathbf{R}$].   Then $\mathscr{B}$ is a base for a topology on X; this topology is called the **metric topology induced by** d.

(e) The **usual topology** on the plane $\mathbf{R}^2$ is the metric topology induced by the function d given by

$$d((x_1, y_1), (x_2, y_2)) = \sqrt{(x_1 - x_2)^2 + (y_1 - y_2)^2}.$$

(f) For $a, b \in \mathbf{R}$, let $[a, b)$ be the set

$$\{x \in \mathbf{R} \mid a \leq x < b\}.$$

Then $\{[a, b) \mid a, b \in \mathbf{R}\}$ is a base for a topology on **R**; this topology is called the **Sorgenfrey topology**, and the set **R** with this topology is called the **Sorgenfrey line**.

4.12   Definition

A topology τ on a set is said to be **second countable** (or: **has a countable base**) iff there exists a base for τ having only countably many members.

4.13   Definition

A topology on a set X is said to be **first countable** iff each point of X has a countable base of neighborhoods.

4.14   Examples

(a) With the usual topology, **R** is both first and second countable.

(b) An uncountable set with the discrete topology is first countable but not second countable.

(c) An uncountable set with the cofinite topology is not first countable.

4.15   Proposition

Every second countable space is first countable.

4.16   Proposition

If (X, τ) is second countable and if $\mathscr{B}$ is any base for τ, then $\mathscr{B}$ has a countable subset that is also a base for τ.

4.17   Definition

Let (X, τ) be a space.   A set $\mathscr{S}$ of subsets of X is said to be a **subbase for** τ iff X together with the set of all intersections of finite nonempty subsets of $\mathscr{S}$ is a base for τ.

4.18   Proposition

Let X be a set and $\mathscr{S}$ be any set of subsets of X.   Then $\mathscr{S}$ is a subbase for *some* topology on X.

4.19   Definition

Let $(T, \leq)$ be any totally ordered set.   For each $t \in T$, let

$$B_t = \{x \in T \mid x < t\},$$

$$A_t = \{x \in T \mid x > t\}$$

("A" for "above" and "B" for "below").   Then the topology on T with subbase

$$\{A_t \mid t \in T\} \cup \{B_t \mid t \in T\}$$

is called the **order topology on** $(T, \leq)$.

4.20   Examples

(a) The order topologies on **R** and **Z** arising from the usual order are the usual topologies (the usual topology on the set of integers **Z** is the discrete topology).

(b) The first infinite ordinal ω with the order topology is discrete and is topologically indistinguishable from **Z** (cf. Definition 4.55). $\omega + 1$ with the order topology is not discrete.

(c) The first uncountable ordinal Ω with the order topology is first countable but not second countable. $\Omega + 1$ with the order topology is neither first nor second countable.

(d) Let $L = [0, 1) \times \Omega$.   Partially order L by taking $(a, b) < (c, d)$ iff $b < d$ or $(b = d$ and $a < c)$.   Then $(L, \leq)$ is totally ordered, and L with the order topology is called the **long line.** L is first countable but is not second countable.

4.21   Definition

Let (X, τ) be a space.   A subset F of X is said to be τ-**closed** (or just **closed**) iff $X - F$ is τ-open.

4.22   Examples

(a) In the discrete topology on a set X, every subset of X is closed.

(b) A subset of $\mathbf{R}$ of the form $[a,\ b] = \{x \in \mathbf{R} \mid a \le x \le b\}$ is called a **closed interval**.   Every closed interval in $\mathbf{R}$ is a closed set.

(c) " Closed " is not the opposite of " open ": A set can be both [as in (a)] or neither: $[a, b)$ is neither open nor closed in the usual topology on $\mathbf{R}$.

4.23   Proposition

(a) If $\mathscr{F}$ is any nonempty set of closed subsets of a space, then $\bigcap \mathscr{F}$ is closed.

(b) The union of a finite set of closed sets is closed.

4.24   Definition

Let (X, τ) be a space, and let $A \subseteq X$.   The **closure of A with respect to** τ is the closed set

$$\mathrm{cl}_\tau A = \bigcap \{F \mid F \supseteq A \text{ and } F \text{ is } \tau\text{-closed}\}.$$

Points of $\mathrm{cl}_\tau A$ are called τ-**adherent points** (or **adherent points**) of A. (When no confusion is likely, mention of τ will be suppressed in this and in similar notations.)

4.25   Definition

Let (X, τ) be a space, and let $A \subseteq X$.   Then x is called a τ-**accumulation point of** A iff $x \in \mathrm{cl}_\tau(A - \{x\})$.

4.26   Proposition

Let (X, τ) be a space and let $A \subseteq X$.   Then

$$\mathrm{cl}_\tau A = \{x \in X \mid \text{every neighborhood of } x \text{ meets } A\}.$$

4.27   Examples

(a) In the usual topology on $\mathbf{R}$,

   (1) $\mathrm{cl}(0, 1) = [0, 1]$.

   (2) $\mathrm{cl}\left\{\dfrac{1}{n} \,\middle|\, n \in \mathbf{N}\right\} = \left\{\dfrac{1}{n} \,\middle|\, n \in \mathbf{N}\right\} \cup \{0\}$, where $\mathbf{N}$ denotes the set of positive integers.

   (3) $\mathrm{cl}\mathbf{Q} = \mathrm{cl}(\mathbf{R} - \mathbf{Q}) = \mathbf{R}$, where $\mathbf{Q}$ denotes the set of rational numbers.

(b) In the discrete topology on a set, each subset is its own closure.

(c) If X has the indiscrete topology, τ, then $\mathrm{cl}_\tau A = X$ whenever $\varnothing \ne A \subseteq X$.

4.28   **Proposition**
Let (X, τ) be a space, and let A and B be subsets of X.   Then
(a) $A \subseteq B$ implies cl $A \subseteq$ cl $B =$ cl(cl B).
(b)   cl$(A \cap B) \subseteq$ (cl A) $\cap$ (cl B).
(c)   cl$(A \cup B) =$ (cl A) $\cup$ (cl B).
(d) A is closed iff cl $A = A$.
(e)   cl $\varnothing = \varnothing$ and cl $X = X$.

4.29   **Definition**
Let (X, τ) be a space, and let $A \subseteq X$.   The **interior of A with respect to** τ is the open set

$$\text{int}_\tau A = \bigcup \{U \mid U \subseteq A \text{ and } U \text{ is } \tau\text{-open}\}.$$

4.30   **Proposition**
Let (X, τ) be a space, and let $A \subseteq X$.   Then
(a)   cl$_\tau(X - A) = X - \text{int}_\tau A$.
(b)   int$_\tau(X - A) = X - \text{cl}_\tau A$.

4.31   **Examples**
(a)   In the usual topology on **R**,
(1)   int$[0, 1] = (0, 1)$.
(2)   int **Q** $= \varnothing$.
(b)   In the discrete topology on a set, each subset is its own interior.
(c)   In the usual topology on the plane,

$$\text{int}\{(x, 0) \mid a < x < b\} = \varnothing.$$

4.32   **Proposition**
Let (X, τ) be a space, and let $A \subseteq X$.   A point x of X belongs to the interior of A iff A is a neighborhood of x.

4.33   **Definition**
Let (X, τ) be a space, and let $A \subseteq X$.   The **frontier of A with respect to** τ is the closed set

$$\text{fr}_\tau A = (\text{cl}_\tau A) \cap (\text{cl}_\tau(X - A)).$$

4.34   **Examples**
(a)   In the usual topology on **R**,
(1)   fr$(0, 1) =$ fr$[0, 1] = \{0, 1\}$.
(2)   fr **Q** $=$ **R**.
(b)   In the discrete topology on any set, the frontier of any subset is empty.
(c)   In the usual topology on the plane,
(1)   fr$\{(x, 0) \mid a \le x \le b\} = \{(x, 0) \mid a \le x \le b\}$.

$$(2) \quad \text{fr}\{(x,y) \mid (x-a)^2 + (y-b)^2 < r\}$$
$$= \{(x,y) \mid (x-a)^2 + (y-b)^2 = r\}, \qquad \text{if } r > 0.$$

4.35   Proposition
Let (X, τ) be a space, and let $A \subseteq X$.   Then
    (a) A is open iff $A \cap \text{fr}_\tau A = \varnothing$.
    (b) A is closed iff $A \supseteq \text{fr}_\tau A$.
    (c) $\text{fr}_\tau A = (\text{cl}_\tau A) - (\text{int}_\tau A)$.

4.36   Definition
Let (X, τ) be a space, and let $A \subseteq X$.   Then A is said to be **nowhere dense in** X iff $\text{int}_\tau \text{cl}_\tau A = \varnothing$.

4.37   Proposition
Let (X, τ) be a space, and let $A \subseteq X$.   If A is either open or closed, then $\text{fr}_\tau A$ is nowhere dense in X.

4.38   Definition
Let (X, τ) be a space.   A subset A of X is said to be **dense in** X iff $\text{cl}_\tau A = X$.

4.39   Proposition
A subset of a space is dense in the space iff it meets every nonempty open subset of the space.

4.40   Definition
A topological space is called **separable** iff it contains a countable dense subset.

4.41   Proposition
Every second countable space is separable.

4.42   Examples
    (a) The Sorgenfrey line is separable but not second countable.
    (b) $\Omega + 1$ with the order topology is not separable.

4.43   Definition
Let (X, τ) be a space, and let Y be a subset of X.   The **relative topology** of τ on Y is the family of sets $\{U \cap Y \mid U \in \tau\}$; Y with the relative topology is called a **subspace** of X.

4.44   Proposition
The relative topology on a subset of a topological space is a topology.

4.45   Definition

A property of topological spaces is called **hereditary** iff every subspace of every space with the property also has the property.

4.46   Examples

(a) Second countability is hereditary; i.e., every subspace of a second countable space is second countable.

(b) Discreteness and indiscreteness are hereditary.

(c) First countability is hereditary.

(d) Separability is not hereditary.

4.47   Proposition

Let (X, τ) be a space, and let (Y, σ) be a subspace.   Then, for $A \subseteq Y$,

(a) $\mathrm{cl}_\sigma A = Y \cap \mathrm{cl}_\tau A$.

(b) $\mathrm{int}_\sigma A \supseteq Y \cap \mathrm{int}_\tau A$.

4.48   Example

Considering $\mathbf{Q}$ as a subspace of $\mathbf{R}$ in the usual topology, $\mathrm{int}_{\mathbf{Q}} \mathbf{Q} = \mathbf{Q}$ but $\mathrm{int}_{\mathbf{R}} \mathbf{Q} = \varnothing$.   Hence proper inclusion may hold in 4.47(b).

4.49   Proposition

Let (X, τ) be a space, and let Y be a dense subset.   For any open subset U of X,

$$\mathrm{cl}_\tau U = \mathrm{cl}_\tau (U \cap Y).$$

4.50   Definition

Let (X, τ) and (Y, σ) be spaces.   A function $f : X \to Y$ is said to be **continuous with respect to τ and σ** iff for every $V \in \sigma$, $f^{-1}[V] \in \tau$; i.e., inverse images of open sets are open.

4.51   Theorem

Let (X, τ) and (Y, σ) be spaces, and let $f : X \to Y$.   The following are equivalent:

(a) f is continuous.

(b) For every σ-closed set B, $f^{-1}[B]$ is τ-closed.

(c) There exists a subbase $\mathscr{S}$ for σ such that for each $S \in \mathscr{S}$, $f^{-1}[S] \in \tau$.

(d) For each $x \in X$ and each σ-neighborhood N of $f(x)$ there exists some τ-neighborhood G of x such that $f[G] \subseteq N$.

(e) For each $A \subseteq X$, $f[\mathrm{cl}_\tau A] \subseteq \mathrm{cl}_\sigma (f[A])$.

(f) For each $B \subseteq Y$, $\mathrm{cl}_\tau f^{-1}[B] \subseteq f^{-1}[\mathrm{cl}_\sigma B]$.

(g) For each $B \subseteq Y$, $f^{-1}[\mathrm{int}_\sigma B] \subseteq \mathrm{int}_\tau (f^{-1}[B])$.

Compare part (d) with the definition of continuity used in calculus.

4.52   Examples
The following functions are continuous:

(a) The identity function from any space X onto itself.

(b) Any inclusion function $i_Y : Y \to X$, where Y is a subspace of X.

(c) Any function $f : X \to Y$, where X has the discrete topology and Y has any topology.

(d) Any function $f : X \to Y$, where X has any topology and Y has the indiscrete topology. [This and the previous part can be used to show that the inclusions in 4.51(e), (f), and (g) may be strict.]

(e) The identity function mapping the Sorgenfrey line onto $\mathbf{R}$, or in general, the identity function $i : (X, \tau) \to (X, \sigma)$ where $\tau \supseteq \sigma$.

(f) Any **constant function**, i.e., any function $c : X \to Y$ where X and Y are any spaces and $c[X]$ is at most a singleton.

(g) The projection functions π_1 and π_2 from $\mathbf{R}^2$ to $\mathbf{R}$.

(h) The function $f : [a, b] \to [c, d]$ defined by

$$f(x) = \begin{cases} c & \text{if } a \leq x \leq c, \\ x & \text{if } c \leq x \leq d, \\ d & \text{if } d \leq x \leq b, \end{cases}$$

where $[c, d] \subsetneq [a, b]$ are considered to be subspaces of $\mathbf{R}$.

(i) The function $g : \mathbf{R} \to S^1$ defined by

$$g(x) = (\cos x, \sin x)$$

where $S^1 = \{(a, b) \mid (a^2 + b^2) = 1\} \subseteq \mathbf{R}^2$.

4.53   Proposition
Let (X, τ) and (Y, σ) be spaces, (Z, ρ) be a subspace of (Y, σ), and $f : X \to Z$. Then $f : (X, \tau) \to (Y, \sigma)$ is continuous iff $f : (X, \tau) \to (Z, \rho)$ is continuous.

4.54   Proposition
The composition of continuous functions is continuous.

4.55   Definition
A continuous function $f : X \to Y$ is called a **homeomorphism** iff there exists a continuous function $g : Y \to X$ such that $f \circ g = 1_Y$ and $g \circ f = 1_X$. Spaces X and Y for which such a homeomorphism exists are said to be **homeomorphic**. A property possessed by some spaces is said to be a **topological invariant** or a **topological property** iff it is preserved under homeomorphisms; i.e., if X has the property and $f : X \to Y$ is a homeomorphism, then Y has the property.

4.56   Examples
The following properties are topological invariants: discreteness, indiscreteness, separability, first countability, second countability, possession of a base every member of which has an empty frontier.

4.57   Theorem

Let $f: (X, \tau) \to (Y, \sigma)$. The following are equivalent:

(a) f is a homeomorphism.

(b) f is one-to-one, onto, continuous, and f^{-1} is continuous.

(c) f is one-to-one, onto, continuous, and for every $V \in \tau, f[V] \in \sigma$.

(d) f is one-to-one, onto, continuous, and for every τ-closed set $G, f[G]$ is σ-closed.

4.58

There is no Cantor–Bernstein theorem for spaces. Let $X = \mathbf{R} \cup \{p\}$ where $p \notin \mathbf{R}$ and $\{p\}$ is open and closed. Then each of X and $\mathbf{R}$ has a homeomorphic copy of the other as a subspace, but X and $\mathbf{R}$ are not homeomorphic. There also exist nonhomeomorphic spaces Z and Y and one-to-one continuous *onto* functions $f: Z \to Y$ and $g: Y \to Z$.

4.59   Definition

A continuous function $f: X \to Y$ is called an **embedding** (of X into Y) iff there exists a continuous function $g: f[X] \to X$ such that $g \circ f = 1_X$, where $f[X]$ has the subspace topology induced from Y.

4.60   Proposition

A function $f: X \to Y$ is an embedding iff f is a homeomorphism from X to $f[X]$.

4.61   Examples

(a) Any homeomorphism is an embedding.

(b) Any inclusion function is an embedding.

(c) The functions $g_p : [0, p) \to S^1$ and $f_p : [0, p) \to \mathbf{R}^2$ defined by $g_p(x) = f_p(x) = (\cos x, \sin x)$ are embeddings iff $p < 2\pi$.

4.62   Definition

A function $f: (X, \tau) \to (Y, \sigma)$ is **open** iff for every τ-open set $A, f[A]$ is σ-open; f is **closed** iff for every τ-closed set $A, f[A]$ is σ-closed.

4.63   Examples

(a) Every homeomorphism is both open and closed.

(b) $\pi_1, \pi_2 : \mathbf{R}^2 \to \mathbf{R}$ are open functions which are not closed.

(c) The function f of 4.52(h) is closed but not open unless $c = d$.

4.64   Proposition

A function $f: X \to Y$ is closed and continuous iff for every $A \subseteq X, f[\operatorname{cl} A] = \operatorname{cl} f[A]$.

Only one of the implications of 4.64 is true if "closed" is replaced by "open" and "cl" by "int".

4.65   **Proposition**
If $f: (X, \tau) \to (Y, \sigma)$ is an open or closed continuous surjection, then σ is the finest topology on Y for which the function f is continuous.

Not only open and closed continuous surjections enjoy this property.   For example, $g_{3\pi}$ of 4.61(c) satisfies the property but is neither open nor closed.

4.66   **Definition**
A continuous function $f: (X, \tau) \to (Y, \sigma)$ is called an **identification map** iff σ is the finest topology on Y for which f is continuous.   If this is the case, then σ is called the **identification topology on Y induced by f**, and (Y, σ) is called an **identification space**.

The reason for this terminology will be made clear later (see 4.73). Note that if $f: X \to Y$ is an identification map and $y \in Y - f[X]$, then $\{y\}$ is open.

4.67   **Examples**
(a) A function $f: X \to Y$, where X has the discrete topology, is an identification map iff Y is discrete.
(b) Any open or closed continuous surjection is an identification map.
(c) The projection maps π_1 and π_2 of $\mathbf{R}^2$ onto $\mathbf{R}$ are identification maps.
(d) The function g_p of 4.61(c) is an identification map iff $p > 2\pi$.

4.68   **Proposition**
A function $f: (X, \tau) \to (Y, \sigma)$ is an identification map iff

$$\sigma = \{U \subseteq Y \mid f^{-1}(U) \in \tau\}.$$

4.69   **Proposition**
A function $f: (X, \tau) \to (Y, \sigma)$ is an identification map iff for every space (Z, μ) and every function $g: (Y, \sigma) \to (Z, \mu)$, g is continuous iff $g \circ f$ is continuous.

4.70   **Proposition**
The composition of identification maps is an identification map.

4.71   **Definition**
Let (X, τ) be a space and R be an equivalence relation on X.   Let τ_R be the identification topology on X/R induced by the natural mapping $\eta : X \to X/R$. Then $(X/R, \tau_R)$ is called the **quotient space of X modulo R**.

Since there is no essential difference between the equivalence relations on X and the decompositions of X into mutually disjoint subsets, quotient spaces of X are sometimes referred to as **decomposition spaces** of X.

4.72   **Examples**

(a) Let $X = [0, 3\pi]$ and let D be the partition

$$\{\{\theta, \theta + 2\pi\} \mid 0 \leq \theta \leq \pi\} \cup \{\{\theta\} \mid \pi < \theta < 2\pi\}.$$

If R is the equivalence relation determined by D, then X/R is homeomorphic to the circle S^1.

(b) Let $X = \{(x, y) \mid x^2 + y^2 \leq 1\}$ be the closed unit disk in $\mathbf{R}^2$. Define an equivalence relation S on X by

(1) $(x, y) \, S \, (x, y)$ for all $(x, y) \in X$ and

(2) $(x, y) \, S \, (-x, -y)$ for all $(x, y) \in X$ such that $x^2 + y^2 = 1$.

The quotient space X/S is called the **projective plane**.

(c) Let $X = [-1, 1] \times [-1, 1] \subseteq \mathbf{R}^2$. Let T be the equivalence relation on X generated by

(1) $(x, y) \, T \, (x, y)$ for all $(x, y) \in X$,

(2) $(x, y) \, T \, (x, -y)$ for all $(x, y) \in X$ with $|y| = 1$, and

(3) $(x, y) \, T \, (-x, y)$ for all $(x, y) \in X$ with $|x| = 1$.

The quotient space X/T is called a **torus**. If condition (3) is deleted, then X/T is homeomorphic to a closed truncated cylinder. If (3) is changed to

(3′) $(x, y) \, T \, (-x, -y)$ for all $(x, y) \in X$ with $|x| = 1$.

then X/T is called the **Klein bottle** (it cannot be embedded in $\mathbf{R}^3$). If (2) is deleted and (3) is changed to (3′), then X/T is called the **Möbius band** (see the frontispiece).

4.73

The following theorem shows that quotient spaces and identification spaces induced by onto functions are topologically indistinguishable.

Theorem

If $f : X \to Y$ is a function, then the following are equivalent:

(a) f is an identification map *onto* Y.

(b) There is an equivalence relation R on X and a homeomorphism $h : Y \to X/R$ such that the diagram

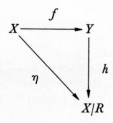

commutes.

This theorem is false if the condition of commutativity is removed.

4.74 **Definition**
A function satisfying the equivalent conditions of Theorem 4.73 is called a **quotient map**.

4.75 **Proposition**
A continuous function is a homeomorphism iff it is an injective quotient map.

4.76 **Definition**
A topological property is called **divisible** iff every quotient space of every space with the property also has the property.

4.77 **Examples**
Separability, discreteness, and indiscreteness are divisible properties, whereas first countability and second countability are not.

4.78 **Theorem**
Let $((X_\alpha, \tau_\alpha))_{\alpha \in A}$ be a family of spaces indexed by a set A, let Y be a set, and for each $\alpha \in A$, let $f_\alpha : X_\alpha \to Y$ be a function. There exists a finest topology σ on Y for which each f_α is continuous. Furthermore, σ is characterized by the following equivalent conditions:

    (a) $\sigma = \{U \subseteq Y \mid \text{for each } \alpha \in A, f_\alpha^{-1}[U] \in \tau_\alpha\}$.

    (b) If (Z, μ) is a space and $g : (Y, \sigma) \to (Z, \mu)$, then g is continuous iff $g \circ f_\alpha$ is continuous for all $\alpha \in A$.

Compare this with 4.68 and 4.69.

4.79 **Definition**
The topology of Theorem 4.78 is called the **fine topology** on Y **induced by** the family $(f_\alpha)_{\alpha \in A}$.

4.80 **Examples**
    (a) The fine topology induced by a single function f is precisely the identification topology induced by f.

    (b) Y has the fine topology with respect to a single *onto* function $f : X \to Y$ iff f is a quotient map.

    (c) Let $\{\tau_\alpha \mid \alpha \in A\}$ be a set of topologies on a set X. Let $\bigwedge\{\tau_\alpha \mid \alpha \in A\}$ be the fine topology on X induced by the identity functions $i_\alpha : (X, \tau_\alpha) \to X$; in the set of all topologies on X (partially ordered by $\subseteq$) this is the infimum of $\{\tau_\alpha \mid \alpha \in A\}$. In general, if $A \neq \varnothing$, then $\bigwedge\{\tau_\alpha \mid \alpha \in A\} = \bigcap\{\tau_\alpha \mid \alpha \in A\}$.

    (d) If $\{(U_\alpha, \tau_\alpha) \mid \alpha \in A\}$ is a set of open subspaces of a space (Y, σ) which covers Y (i.e., $\bigcup\{U_\alpha \mid \alpha \in A\} = Y$), then σ is the fine topology on Y induced by the inclusions $i_\alpha : U_\alpha \to Y$.

4.81   Definition

Let $((X_\alpha, \tau_\alpha))_{\alpha \in A}$ be a family of spaces, and let $Y = \sum (X_\alpha)$. Then Y provided with the fine topology induced by the injections $\mu_\alpha : X_\alpha \to Y$ is called the **topological sum** or **disjoint topological union** of $((X_\alpha, \tau_\alpha))_{\alpha \in A}$.

4.82   Examples

(a) Every discrete space is the topological sum of singleton spaces.

(b) The space $\{(x, m) \mid x \in \mathbf{R}, \ m \in \omega\} \subseteq \mathbf{R}^2$ is homeomorphic to a topological sum of $\aleph_0$ copies of $\mathbf{R}$.

(c) $\{(a, b) \in \mathbf{R}^2 \mid a = 0 \text{ or } b = 0\}$ is not homeomorphic to a topological sum of two copies of $\mathbf{R}$.

4.83   Theorem

Let $\sum X_\alpha$ be the topological sum of spaces X_α, $\alpha \in A$, with injections $\mu_\alpha : X_\alpha \to \sum X_\alpha$.

(a) If Z is any space, and if for every $\alpha \in A$, $f_\alpha : X_\alpha \to Z$ is continuous, then there exists a unique continuous function $f : \sum X_\alpha \to Z$ such that for every $\alpha \in A$, the diagram

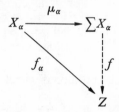

commutes.

(b) The above property characterizes the topological sum; i.e., if W is a space and if for every $\alpha \in A$, $\nu_\alpha : X_\alpha \to W$ is continuous and if for every space Z together with continuous functions $f_\alpha : X_\alpha \to Z$ there exists a unique continuous function $f : W \to Z$ such that for every $\alpha \in A$, $f \circ \nu_\alpha = f_\alpha$, then there exists a homeomorphism $h : \sum X_\alpha \to W$ such that for every $\alpha \in A$, the diagram

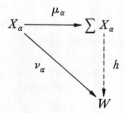

commutes.

Compare this theorem with 2.58.

4.84   **Definition**
A topological property is called **summable** iff it is preserved by topological sums.

4.85   **Proposition**
First countability and discreteness are summable; second countability and separability are not.

4.86

By interchanging the domains and ranges of the functions in question in the previous theorem, we "dualize" the notion of fine topology and topological sum.

Theorem
Let X be a set, $((Y_\alpha, \sigma_\alpha))_{\alpha \in A}$ a family of spaces indexed by a set A, and for every $\alpha \in A$, let $g_\alpha : X \to Y_\alpha$ be a function. Then there exists a coarsest topology τ on X for which each g_α is continuous. Furthermore, τ is characterized by the following equivalent conditions:
(a) $\{g_\alpha^{-1}[U] \mid U \in \sigma_\alpha, \alpha \in A\}$ is a subbase for τ.
(b) If (Z, μ) is a space and $f : (Z, \mu) \to (X, \tau)$, then f is continuous iff $g_\alpha \circ f$ is continuous for all $\alpha \in A$.

4.87   **Definition**
The topology τ of Theorem 4.86 is called the **coarse topology** on X **induced by** the family $(g_\alpha)_{\alpha \in A}$.

4.88   **Examples**
(a) X has the coarse topology induced by a single one-to-one function $f : X \to Y$ iff f is an embedding.
(b) Let $\{\tau_\alpha \mid \alpha \in A\}$ be a set of topologies on a set X. Let $\bigvee\{\tau_\alpha \mid \alpha \in A\}$ be the coarse topology on X induced by the identity functions $i_\alpha : X \to (X, \tau_\alpha)$. In the set of all topologies on X partially ordered by $\subseteq$, $\bigvee\{\tau_\alpha \mid \alpha \in A\}$ is the supremum of $\{\tau_\alpha \mid \alpha \in A\}$. In general $\bigvee\{\tau_\alpha \mid \alpha \in A\} \neq \bigcup\{\tau_\alpha \mid \alpha \in A\}$, even when $A \neq \varnothing$.
(c) The usual topology on $\mathbf{R}^2$ is the coarse topology induced by the projections $\pi_1, \pi_2 : \mathbf{R}^2 \to \mathbf{R}$.
(d) Let (X, τ) and (Y, σ) be spaces. Let $\tau \times \sigma$ be the topology on $X \times Y$ with base $\{U \times V \mid U \in \tau, V \in \sigma\}$. Then $\tau \times \sigma$ is the coarse topology on $X \times Y$ induced by the projections $\pi_1 : X \times Y \to X$ and $\pi_2 : X \times Y \to Y$.

4.89   **Definition**
Let $((Y_\alpha, \sigma_\alpha))_{\alpha \in A}$ be a family of spaces, and let $X = \Pi(Y_\alpha)$. The coarse topology on X induced by the projections $\pi_\alpha : X \to Y_\alpha$ is called the **product topology** on X, and X equipped with this topology is called the **product**

space of the **coordinate spaces** Y_α, $\alpha \in A$. If the coordinate spaces are all the same, then the product space is denoted by Y^A.

4.90   *Examples*
(a) $\mathbf{R}^n$ has its usual meaning (i.e., euclidean n-dimensional space).

(b) For each $\alpha \in A$, let $(Y_\alpha, \sigma_\alpha)$ be a space. The set consisting of all sets of the form ΠU_α, where each $U_\alpha \in \sigma_\alpha$, is a base for a topology on ΠY_α. This topology is called the **box topology**. In general, it is different from the product topology on ΠY_α.

The box topology and the product topology do coincide, however, if A is finite or if σ_α is indiscrete for all but a finite number of the α.

4.91   *Definition*
Let D be a two-element discrete space. D^ω is said to be the **Cantor discontinuum**.

4.92   *Definition*
For each positive integer n, let

$$F_n = [0, 1] - \left(\left(\frac{1}{3^n}, \frac{2}{3^n} \right) \cup \left(\frac{3}{3^n}, \frac{4}{3^n} \right) \cup \cdots \cup \left(\frac{3^n - 2}{3^n}, \frac{3^n - 1}{3^n} \right) \right).$$

Then $F = \bigcap \{F_n \mid n \in \mathbf{N}\}$ with the relative topology induced from $\mathbf{R}$ is called the **Cantor middle third space**.

4.93   *Examples*
(a) The Cantor discontinuum is homeomorphic to the Cantor middle third space. Either is called the **Cantor space**.

(b) Among the remarkable properties of the Cantor space we find the following:

    (1) The Cantor space F is uncountable but there are only countably many end points of "removed intervals", so that "most" of the points of F are not end points of removed intervals.

    (2) No subset of F with more than one point is connected (see 4.138), so F contains no intervals.

    (3) F is compact (see 4.153).

    (4) If $x \in F$, then x is an accumulation point of F.

(c) The Cantor space provides the basis for a great many surprising examples in analysis. For example, it can be used to construct

    (1) An uncountable set of Lebesgue measure zero.

    (2) A continuous strictly increasing function that has derivative zero almost everywhere.

    (3) A nowhere dense perfect set of measure zero.

    (4) (by a suitable change in construction) a nowhere dense perfect set of positive measure.

4.94   ***Theorem***

Let ΠX_α be the topological product of spaces X_α, $\alpha \in A$, with projections $\pi_\alpha : \Pi X_\alpha \to X_\alpha$.

(a) If Z is any space and if for every $\alpha \in A$, $g_\alpha : Z \to X_\alpha$ is continuous, then there exists a unique continuous function $g : Z \to \Pi X_\alpha$ such that for every $\alpha \in A$, the diagram

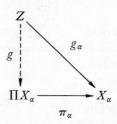

commutes.

(b) The above property characterizes the topological product; i.e., if W is a space and if for every $\alpha \in A$, $\rho_\alpha : W \to X_\alpha$ is continuous and if for every space Z together with continuous functions $g_\alpha : Z \to X_\alpha$ there exists a unique continuous function $g : Z \to W$ such that for every $\alpha \in A$, $\rho_\alpha \circ g = g_\alpha$, then there exists a homeomorphism $h : W \to \Pi X_\alpha$ such that for every $\alpha \in A$ the diagram

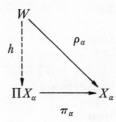

commutes.

4.95   ***Example***

If $(X_\alpha)_{\alpha \in A}$ is a family of topological spaces and if Y is the *set* ΠX_α with the box topology, then Y need not satisfy the preceding theorem. Since in almost every setting [e.g., set theory (see 2.56), group theory, and ring theory] products are characterized essentially by this theorem, this is a significant reason for not taking the box topology to be the usual topology on a product of spaces. There are other reasons (see 4.160).

4.96   ***Definition***

Let $(X_\alpha)_{\alpha \in A}$ be a family of spaces and let $F = \{f_\alpha \mid \alpha \in A\}$ be a set of continuous functions with $f_\alpha : X \to X_\alpha$. The unique continuous function (guaranteed by Theorem 4.94) $e : X \to \Pi X_\alpha$ such that $\pi_\alpha \circ e = f_\alpha$ for every $\alpha \in A$ is called the **evaluation map** from X to ΠX_α (with respect to F).

4.97   **Definition**
A topological property is called **productive** iff it is preserved by topological products.

4.98   **Proposition**
Among the properties discrete, indiscrete, separable, first countable, and second countable, only indiscrete is productive.

4.99   **Proposition**
If each space X_α is nonempty, then for each α there exists an embedding $e_\alpha : X_\alpha \to \Pi X_\alpha$ such that $\pi_\alpha \circ e_\alpha = 1_{X_\alpha}$; hence in this case each projection is surjective.

4.100   **Proposition**
Every projection $\pi_\alpha : \Pi X_\alpha \to X_\alpha$ is an open function but is not necessarily a closed function.

4.101   **Corollary**
If each X_α is nonempty, then every projection $\pi_\alpha : \Pi X_\alpha \to X_\alpha$ is a quotient map.

4.102   **Definition**
A space X is said to be T_1 iff every singleton subset of X is closed.

4.103   **Proposition**
Let $f : X \to Y$ be a quotient map.   Then Y is T_1 iff for each $x \in X, f^{-1}[f(x)]$ is closed; i.e., Y is T_1 iff all the equivalence classes corresponding to the quotient map are closed.

4.104   **Proposition**
"T_1" is productive, hereditary, and summable, but it is not divisible.

4.105   **Proposition**
For every set X, there exists a coarsest T_1 topology on X.

4.106   **Definition**
If a set F of continuous functions, each with domain X, has the property that for every closed $C \subseteq X$ and $x \notin C$, there exists some $f \in F$ such that $f(x) \notin \operatorname{cl} f[C]$, then F is said to **distinguish points and closed sets**.

4.107   **Theorem (Embedding Theorem)**
Let X be a T_1 space, let $(X_\alpha)_{\alpha \in A}$ be a family of spaces, and let $F = \{f_\alpha \mid \alpha \in A\}$ be a set of continuous functions $f_\alpha : X \to X_\alpha$.   If F distinguishes points and closed sets, then the evaluation map with respect to F, $e : X \to \Pi X_\alpha$, is an embedding.

4.108 **Definition**
A set $\mathscr{F}$ of subsets of a set X is called a **filter on** X provided that the following conditions hold:
(a) $\varnothing \neq \mathscr{F}$.
(b) $\varnothing \notin \mathscr{F}$.
(c) If $F_1, F_2 \in \mathscr{F}$, then $F_1 \cap F_2 \in \mathscr{F}$.
(d) If $F \in \mathscr{F}$ and $F \subseteq G \subseteq X$, then $G \in \mathscr{F}$.
A filter on X which is not contained in any other filter is called a **maximal filter**.

4.109 **Examples**
(a) In any space, the set $\mathscr{N}_x$ of all neighborhoods of a fixed point x is a filter.
(b) In any infinite set, the set of all subsets with finite complement is a filter.
(c) In any set with cardinality greater than or equal to $\aleph_\alpha$, the set of all subsets with complement having cardinality less than $\aleph_\alpha$ is a filter.
(d) In any set, the set $\mathscr{F}_x$ of all subsets of X containing a fixed point $x \in X$ is a maximal filter on X.

4.110 **Definition**
Let X be a space, $\mathscr{F}$ a filter on X, and x a point of X. Then we say that $\mathscr{F}$ **converges to** x (or x is a **limit** of $\mathscr{F}$) iff each neighborhood of x belongs to $\mathscr{F}$; x is called a **cluster point** of $\mathscr{F}$ iff each neighborhood of x meets each element of $\mathscr{F}$.

4.111 **Examples**
(a) The filter $\mathscr{N}_x$ of all neighborhoods of a point x converges to x and has x as a cluster point. It may converge to other points and have other cluster points. In **R**, each $\mathscr{N}_x$ converges only to x.
(b) In **R**, the "cofinite" filter [4.109(b)] and the "cocountable" filter [4.109(c) with $\alpha = 1$] have every point of **R** as cluster points, but neither converges.
(c) For any space X and any $x \in X$, the filter $\mathscr{F}_x$ of all subsets containing x converges to x.

4.112 **Definition**
A set $\mathscr{F}$ of sets is said to have the **finite intersection property** (**f i p**) iff every finite subset of $\mathscr{F}$ has a nonempty intersection.

4.113 **Proposition**
If X is a nonempty set and if $\mathscr{S}$ is a set of subsets of X with f i p, then there exists a smallest filter containing $\mathscr{S}$.

4.114   Definition

The unique smallest filter containing a set $\mathscr{S}$ of subsets with f i p is called the **filter generated by** $\mathscr{S}$.

4.115   Proposition

A point x is a cluster point of a filter $\mathscr{F}$ iff there exists a filter containing $\mathscr{F}$ that converges to x.

4.116   Examples

(a) The set of all subsets of **R** of the form $\{x \in \mathbf{R} \mid x > r\}$ for $r \in \mathbf{R}$ has f i p, and hence generates a filter. This filter does not converge and it has no cluster points.

(b) Let $(s_n)_{n \in \omega}$ be a sequence in **R**. Let $\mathscr{T}_s$ be the filter generated by the set $\{\{s_n \mid n \geq m\} \mid m \in \omega\}$ of **tails** of (s_n). Then $\mathscr{T}_s$ converges iff (s_n) does (see 3.37), and if they converge, then both have the same limit.

In general, we will say that a sequence (s_n) in a space X **converges** iff the filter generated by the set of tails converges.

4.117   Proposition

Let $\mathscr{F}$ be a filter on a topological space X. If $\mathscr{F}$ converges to x, then x is also a cluster point of $\mathscr{F}$.

4.118   Theorem

A function $f : X \to Y$ is continuous iff the following condition holds: For all $x \in X$ and for each filter $\mathscr{F}$ converging to x, the filter generated by $\{f[F] \mid F \in \mathscr{F}\}$ converges to $f(x)$.

4.119   Example

If (s_n) is a sequence in X that converges to x, and if $f : X \to Y$ is continuous, then the sequence $(f(s_n))_{n \in \omega}$ converges to $f(x)$. However, this does not characterize continuity: the function $f : (\Omega + 1) \to \mathbf{R}$ given by $f[\Omega] = \{0\}$, $f(\Omega) = 1$ is not continuous, even though it does preserve sequential convergence.

Next we generalize the concept of sequence in such a way that convergence here will characterize continuity, just as convergence of filters does.

4.120   Definition

(a) A **directed set** is a quasi-ordered set $(D, \leq)$ such that for all x, $y \in D$, there exists a $z \in D$ with $x \leq z$ and $y \leq z$.

(b) If X is any set, a **net in** X is a function from a directed set into X.

(c) Let X be a topological space and let $s : D \to X$ be a net in X. Then we say that s **converges to** $x \in X$ (or x is a **limit** of s) provided that for each neighborhood U of x, there exists a $d \in D$ such that $s(d') \in U$ whenever $d' \geq d$ (i.e., the net is **ultimately in every neighborhood of** x).

(d) If $s : D \to X$ is a net in X, the **filter associated with** s is the filter generated by

$$\{\{s(d) \mid d \geq d'\} \mid d' \in D\}.$$

4.121   Theorem
Let A be a subset of a space X.   Then

$$\operatorname{cl} A = \{x \in X \mid \text{for some net } s \text{ in } X \text{ whose range is contained}$$
$$\text{in } A, s \text{ converges to } x\}.$$

4.122   Proposition
Let $\mathscr{F}$ be a maximal filter on X.   Then there exists a net $s : D \to X$ such that $\mathscr{F}$ is the filter associated with s.

4.123   Proposition
If $\mathscr{F}$ is the filter associated with a net s in a space X, then $\mathscr{F}$ converges to $x \in X$ iff s converges to x.

4.124   Theorem
Let X and Y be spaces, and let $f : X \to Y$.   Then f is continuous iff for each $x \in X$ and each net s converging to x, the net $f \circ s$ converges to $f(x)$.

4.125   Theorem
Every filter on a set is contained in a maximal filter.

4.126   Proposition
Let $\mathscr{F}$ be a filter on a set.   The following are equivalent:
  (a) $\mathscr{F}$ is a maximal filter.
  (b) If G meets every element of $\mathscr{F}$, then $G \in \mathscr{F}$.
  (c) If $H \notin \mathscr{F}$ and $K \notin \mathscr{F}$, then $H \cup K \notin \mathscr{F}$.

4.127   Proposition
Let $\mathscr{F}$ be a maximal filter on a space.   Then $\mathscr{F}$ converges to a point x iff x is a cluster point of $\mathscr{F}$.

4.128   Theorem
A filter $\mathscr{F}$ on a product $\Pi(X_\alpha)_{\alpha \in A}$ of spaces converges to a point y iff for each $\alpha \in A$, the filter $\{\pi_\alpha[F] \mid F \in \mathscr{F}\}$ converges to $\pi_\alpha(y)$.

4.129   Proposition
Let X be a space.   The following are equivalent:
  (a) Every filter on X that converges has a unique limit.

(b) If $x, y \in X$ with $x \neq y$, then there exist disjoint neighborhoods of x and y.

4.130   Definition
A space X is said to be **Hausdorff** (or T_2) iff every pair of distinct points of X have disjoint neighborhoods.

4.131   Examples
(a) An infinite set with the cofinite topology is T_1 but not Hausdorff.
(b) **R**, **Q**, and **N** are Hausdorff.

4.132   Proposition
Any topology finer than a T_2 topology is T_2.

4.133   Proposition
A space X is Hausdorff iff the set $\{(x, x) \mid x \in X\}$ is closed in $X \times X$.

4.134   Theorem
Let X be a space and let Y be a Hausdorff space. Let $f: X \to Y$ and $g: X \to Y$ be continuous. Then $\{x \in X \mid f(x) = g(x)\}$ is closed.

4.135   Corollary
If X and Y are spaces with Y Hausdorff, and if $f: X \to Y$, $g: X \to Y$ are continuous and agree on a dense subset of X, then $f = g$.

4.136   Theorem
"Hausdorff" is productive, hereditary, and summable, but is not divisible.

4.137   Proposition
Every Hausdorff space is T_1.

4.138   Definition
A topological space is said to be **connected** iff it cannot be expressed as a union of two disjoint nonempty open subsets. A subset of a space is said to be **connected** iff it is connected as a topological space with the relative topology.

4.139   Examples
(a) **R** and all its intervals (open, closed, or half open and half closed; bounded or unbounded) are connected. These are the only connected subsets of **R**.
(b) **Q** and **N** are not connected.
(c) $\mathbf{R}^2$ is connected.
(d) The long line is connected.

(e) Any set with the indiscrete topology is connected, and no set with more than one point and the discrete topology is connected.

4.140   **Proposition**
If U is a connected subset of a topological space and if $U \subseteq V \subseteq \mathrm{cl}\ U$, then V is connected.

4.141   **Proposition**
Let $\{Y_\alpha \mid \alpha \in A\}$ be a set of connected subsets of X such that for all $\alpha,\ \beta \in A$, $Y_\alpha \cap Y_\beta \neq \varnothing$. Then $\bigcup \{Y_\alpha \mid \alpha \in A\}$ is connected.

4.142   **Proposition**
Every connected subset of a topological space is contained in a maximal connected subset.

4.143   **Definition**
A maximal connected subset of a topological space is called a **component**.

4.144   **Proposition**
Let X be a space. Then
   (a) Components of X are closed.
   (b) The components form a partition of X.

4.145   **Theorem**
If X is connected and f is a continuous function from X onto Y, then Y is connected.

4.146   **Corollary**
Any topology coarser than a connected topology is connected.

4.147   **Examples**
The following applications to calculus follow easily from the preceding theorem.
   (a) *Intermediate Value Theorem.* If $f : \mathbf{R} \to \mathbf{R}$ is continuous on an interval containing a and b, and if $f(a) \le c \le f(b)$, then there exists some d between a and b such that $f(d) = c$.
   (b) *Bolzano's Theorem.* If $f : [a, b] \to \mathbf{R}$ is continuous and if $f(a)$ and $f(b)$ have opposite signs, then there is some $c \in [a, b]$ such that $f(c) = 0$.

4.148   **Proposition**
X is connected iff the two-point discrete space is not a quotient space of X.

4.149   **Theorem**
$\mathbf{R}$ and $\mathbf{R}^2$ are not homeomorphic.

4.150   Theorem

A space X is connected iff for every $x, y \in X$, there is a finite sequence $\{G_i \mid 1 \leq i \leq n\}$ of connected subsets of X such that $x \in G_1$, $y \in G_n$, and G_i meets G_{i+1} for each i such that $1 \leq i < n$.

4.151   Theorem

Any product of connected spaces is connected.

4.152   Proposition

Connectedness is neither summable nor hereditary.

4.153   Definition

Let X be a space. A set $\mathcal{F}$ of open subsets of X for which $\bigcup \mathcal{F} = X$ is called an **open cover** of X.

If $\mathcal{F}$ is an open cover of X, a **subcover** is a subset $\mathcal{D}$ of $\mathcal{F}$ whose union is X. $\mathcal{D}$ is called a **finite subcover** if $\mathcal{D}$ consists of only finitely many sets.

A space X is said to be **compact** iff every open cover of X has a finite subcover.

4.154   Examples

(a) Any closed interval in $\mathbf{R}$ is compact.

(b) (a, b) and $\{x \in \mathbf{R} \mid x \geq a\}$ are not compact.

(c) A subset of $\mathbf{R}$ is compact iff it is closed and bounded.

(d) $\Omega + 1$ with the order topology is compact. However, there is an open cover of Ω which does not even have a *countable* subcover.

4.155   Theorem

Let X be a space. The following are equivalent:

(a) X is compact.

(b) If $\mathcal{H}$ is a set of closed subsets of X with f i p, then $\bigcap \mathcal{H} \neq \emptyset$.

(c) Every filter on X has a cluster point.

(d) Every maximal filter on X has a limit (one or more).

4.156   Theorem

Compactness is closed-hereditary (i.e., every closed subspace of a compact space is compact) and is preserved by continuous functions, but is neither hereditary nor summable.

4.157   Examples

The following applications to calculus follow from the preceding theorem.

(a) *Extreme Value Theorem.* If $f : [a, b] \to \mathbf{R}$ is continuous, then f attains an absolute maximum and an absolute minimum on $[a, b]$.

(b) *Mean Value Theorem for Integrals.* If $f : [a, b] \to \mathbf{R}$ is continuous, then there is some $c \in [a, b]$ such that $\int_a^b f = (b - a) f(c)$.

PROOF Let M and m be the absolute maximum and absolute minimum, respectively, for f. Then $m \leq f(x) \leq M$ for each $x \in [a, b]$, so that since integration is order-preserving, we have

$$\int_a^b m \leq \int_a^b f \leq \int_a^b M$$

or

$$m(b - a) \leq \int_a^b f \leq M(b - a).$$

But $m = f(x_1)$ and $M = f(x_2)$ for some x_1, $x_2 \in [a, b]$. Thus we have the desired result by the intermediate value theorem [4.147(a)].

(c) *Rolle's Theorem.* If $f: [a, b] \to \mathbf{R}$ is continuous, $f(a) = f(b)$, and the derivative f' exists on (a, b), then there is some $c \in (a, b)$ such that $f'(c) = 0$.

PROOF Since $f(a) = f(b)$, there is some $c \in (a, b)$ at which f takes on its absolute maximum or its absolute minimum. This is a relative extremum for $f | (a, b)$, so that $f'(c) = 0$.

(d) *Mean Value Theorem for Derivatives.* If $f: [a, b] \to \mathbf{R}$ is continuous and f' exists on (a, b), then there is some $c \in (a, b)$ such that

$$f'(c) = \frac{f(b) - f(a)}{b - a}.$$

PROOF Let $g(x) = f(x)(b - a) - x(f(b) - f(a))$. Apply Rolle's Theorem to $g: [a, b] \to \mathbf{R}$.

4.158 Theorem (Tychonoff Product Theorem)
Compactness is productive.

4.159 Corollary
The Cantor discontinuum D^ω is compact.

4.160 Example
For each $n \in \omega$, let D_n be a two-element discrete space and let X be the *set* $\Pi(D_n)_\omega$ with the box topology. Each D_n is compact, while X is an infinite discrete space and is thus not compact. This is another reason that the usual product topology is favored over the box topology.

4.161 Theorem
If X is compact and Y is any space, then the projection $\pi_2: X \times Y \to Y$ is a closed function.

4.162 Proposition
Every compact subset of a Hausdorff space is closed.

4.163   **Proposition**
Let K be a compact subset of a space X, and L be a compact subset of Y. If U is any open set in $X \times Y$ containing $K \times L$, then there exist open sets $V \supseteq K$ and $W \supseteq L$ such that $K \times L \subseteq V \times W \subseteq U$.

4.164   **Definition**
A space X is said to be **regular** iff for every closed set F and point $x \in X - F$, there exist disjoint open sets U and V with $x \in U$ and $F \subseteq V$. A regular T_1 space is called a T_3 space.

4.165   **Proposition**
A space X is regular iff every point of X has a base of neighborhoods consisting of closed sets.

4.166   **Examples**
(a) Any set with more than one point having the indiscrete topology is regular but not T_3.
(b) Let X be the set of real numbers and let τ be the topology on X having as a subbase the usual topology together with the one set $\{x \in X \mid x \neq 1/n$ for all $n \in \mathbf{N}\}$. Then (X, τ) is Hausdorff but not regular.

4.167   **Theorem**
Regularity is productive, hereditary, and summable, but not divisible.

4.168   **Definition**
A space X is said to be **completely regular** iff whenever F is a nonempty closed subset of X and $x \in X - F$, there exists a continuous function $f : X \to [0, 1]$ for which $f(x) = 0$ and $f[F] = \{1\}$.

4.169   **Examples**
(a) $\mathbf{R}$ and $\mathbf{Q}$ are completely regular.
(b) No infinite set with the cofinite topology is completely regular. Indeed, every completely regular space is regular.
(c) There is an example of a regular T_2 space that is not completely regular (indeed, on which every continuous real-valued function is *constant*), but its construction is very difficult.

4.170   **Theorem**
Complete regularity is productive, hereditary, and summable, but is not divisible.

4.171   **Definition**
Let X be a space. A **compactification** of X is a pair (K, f), where K is a compact space and $f : X \to K$ is an embedding whose image is dense in K. The compactification (K, f) is called **Hausdorff** iff K is Hausdorff.

4.172   *Examples*

(a) [0, 1] together with the inclusion function of (0, 1) into [0, 1] is a compactification of (0, 1).

(b) [0, $\pi/2$] together with the function $x \mapsto \arctan x$ is a compactification of $\{r \in \mathbf{R} \mid r \geq 0\}$.

(c) $\omega + 1$ with the order topology is a compactification of ω with the order topology.

4.173   *Theorem*

Let X be a completely regular Hausdorff space. Then X has a Hausdorff compactification $(\beta X, h)$ with the following equivalent properties:

(a) If $f : X \to [0, 1]$ is continuous, then there exists a unique continuous function $\bar{f} : \beta X \to [0, 1]$ such that the diagram

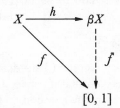

commutes.

(b) If Y is a compact Hausdorff space, and if $g : X \to Y$ is continuous, then there exists a unique continuous function $\bar{g} : \beta X \to Y$ such that the diagram

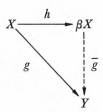

commutes.

4.174   *Definition*

A compactification $(\beta X, h)$ satisfying Theorem 4.173 is called the **Stone–Čech compactification of** X.

The next proposition shows why "the" in the above definition is appropriate.

4.175   Proposition

If (Y, h) and (Y', h') are Stone–Čech compactifications of X, then there exists a homeomorphism $f: Y \to Y'$ such that the diagram

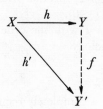

commutes.

4.176   Definition

A space X is said to be **normal** iff whenever C and D are disjoint closed subsets of X, there exist disjoint open sets U and V with $C \subseteq U$ and $D \subseteq V$.

4.177   Theorem

    (a) Every compact Hausdorff space is normal.
    (b) Every normal T_1 space is regular and Hausdorff.
    (c) Every regular second countable space is normal.

4.178   Theorem (Urysohn's Lemma)

If C and D are disjoint closed subsets of a normal space X, then there is a continuous function $f: X \to [0, 1]$ such that $f[C] \subseteq \{0\}$ and $f[D] \subseteq \{1\}$.

4.179   Corollary

Every normal T_1 space is completely regular.

4.180   Corollary

Every compact Hausdorff space is completely regular.

4.181   Corollary

Let X be a Hausdorff space. The following are equivalent:
    (a) X is completely regular.
    (b) X has a Hausdorff compactification.
    (c) X can be embedded in a compact Hausdorff space.

4.182   Example

The space $[(\Omega + 1) \times (\omega + 1)] - \{(\Omega, \omega)\}$ is completely regular and Hausdorff but is not normal. This also shows that normality is not hereditary. (The space $(\Omega + 1) \times (\omega + 1)$ is called the **Tychonoff plank**.)

4.183   **Theorem**
Normality is summable and closed hereditary but is not productive or divisible.

4.184   **Theorem (Tietze Extension Theorem)**
A space X is normal iff every bounded continuous real-valued function defined on a closed subset of X has a continuous extension to all of X.

4.185   **Definition**
A topological space is said to be **locally compact** iff each of its points has a compact neighborhood.

4.186   **Examples**
    (a) Every compact space and every discrete space is locally compact.
    (b) For each $n \in \omega$, $\mathbf{R}^n$ is locally compact.
    (c) Every open subspace of a compact T_2 space is locally compact.
    (d) $\mathbf{Q}$ is not locally compact.

4.187   **Theorem**
Local compactness
    (a) is preserved by *open* continuous functions (and is thus a topological invariant) but is not preserved by all continuous functions,
    (b) is closed hereditary but is not hereditary,
    (c) is not productive but is preserved by finite products, and
    (d) is summable.

4.188   **Proposition**
Every first countable Hausdorff space is a quotient of a locally compact space.

4.189   **Corollary**
Local compactness is not divisible.

4.190   **Theorem**
Every locally compact Hausdorff space is completely regular.

4.191   **Corollary**
A Hausdorff space is locally compact iff each of its points has a neighborhood base consisting of compact sets.

4.192   **Theorem**
A Hausdorff space is locally compact iff for each of its Hausdorff compactifications (K, f), f is an open function.

4.193   Theorem

A space X has a Hausdorff compactification $(\alpha X, f)$ such that $\alpha X - f[X]$ has at most one point iff X is locally compact and Hausdorff. If X has such a compactification, it is unique in the following sense:

If (K, g) is a Hausdorff compactification such that $K - g[X]$ has at most one point, then there exists a homeomorphism $h : \alpha X \to K$ such that the diagram

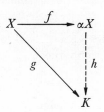

commutes.

4.194   Definition

The compactification $(\alpha X, f)$ of Theorem 4.193 is called the **Alexandroff one-point compactification of** X.

4.195   Examples

(a) $\omega + 1$ with the order topology is (homeomorphic with) the one-point compactification of ω.

(b) S^1 is (homeomorphic with) the one-point compactification of **R**.

(c) If for each $n \in \omega$,

$$S^n = \{(x_1, \ldots, x_{n+1}) \mid x_1{}^2 + \cdots + x_{n+1}{}^2 = 1\} \subseteq \mathbf{R}^{n+1},$$

then S^n is (homeomorphic with) the one-point compactification of $\mathbf{R}^n$.

(d) The one-point compactification of the long line L is just $L \cup \{(1, \Omega)\}$ with the order topology.

4.196   Definition

A space X is said to be a **Baire space** iff the intersection of any countable set of open dense subsets of X is dense in X.

4.197   Proposition

A space X is a Baire space iff no nonempty open subset of X is a countable union of nowhere dense subsets of X.

4.198   Example

The space of rational numbers, $\mathbf{Q}$, is not a Baire space, but for each $n \in \omega$, $\mathbf{R}^n$ is a Baire space (cf. Theorem 4.200).

4.199   Proposition

If $f: X \to Y$ is continuous and open, with $f[X]$ dense in Y, and if X is a Baire space, then Y is a Baire space. Hence "Baire" is a topological property.

4.200   Theorem (Baire Category Theorem, I)

Each locally compact Hausdorff space is a Baire space.

4.201   Definition

(a) Let X be a set. A **metric** on X is a function $d: X \times X \to \mathbf{R}$ such that for each $x, y, z \in X$,

     (1) $d(x, y) = 0$ iff $x = y$,
     (2) $d(x, y) = d(y, x) \geq 0$, and
     (3) $d(x, y) + d(y, z) \geq d(x, z)$.

If d is a metric on X, then (X, d) is called a **metric space**.

(b) If (X, d) is a metric space and if $c \in X$ and $0 < r \in \mathbf{R}$, then

$$B_d(c, r) = \{x \in X \mid d(x, c) < r\}$$

is the **open d-ball with radius r and center c**. The topology for which $\{B_d(c, r) \mid c \in X, 0 < r \in \mathbf{R}\}$ is a base is called the **metric topology induced by d** and is denoted by τ_d.

(c) If (X, τ) is a topological space, and if there exists a metric d on X such that τ is the metric topology induced by d, then (X, τ) is said to be **metrizable**.

(d) If (X, d) is a metric space and $A \subseteq X$, then

$$d(x, A) = \inf \{d(x, y) \mid y \in A\}$$

is called the **distance from x to A**.

4.202   Examples

(a) $\mathbf{R}^n$ is metrizable for every $n \in \omega$.
(b) Every discrete space is metrizable.
(c) Let $H = \{s: \omega \to \mathbf{R} \mid \sum_{i \in \omega} (s(i))^2 \text{ converges}\}$; for $s, t \in H$, define $d(s, t) = \sqrt{\sum_{i \in \omega} (s(i) - t(i))^2}$. Then d is a metric on H. (H, τ_d) is called **Hilbert space**.

4.203   Proposition

A subset S of a set X is closed in the metric topology on (X, d) iff

$$S = \{x \in X \mid d(x, S) = 0\}.$$

4.204   Theorem

Every metrizable space is normal, first countable, and Hausdorff.

4.205   *Examples*
(a) $\Omega + 1$ with the order topology is not metrizable.
(b) Ω with the order topology is not metrizable, even though it is normal, first countable, and Hausdorff.

4.206   *Theorem*
Metrizability is hereditary and summable, but is neither productive nor divisible.

4.207   *Lemma*
Let (X, d) be a metric space.   Then $e : X \times X \rightarrow \mathbf{R}$ defined by

$$e(a, b) = \frac{d(a, b)}{1 + d(a, b)}$$

is a metric, and $\tau_e = \tau_d$.

4.208   *Theorem*
Every countable (finite or infinite) product of metrizable spaces is metrizable.

4.209   *Examples*
(a) The metrizable space $[0, 1]^\omega$ is called the **Hilbert cube**.   It can be embedded in Hilbert space.
[The next two examples are included for illustration.   Part (c) is difficult to prove, and is beyond the scope of this presentation.]
(b) The metrizable space ω^ω is homeomorphic to the space $\mathbf{R} - \mathbf{Q}$ of irrational numbers.
(c) The metrizable space $\mathbf{R}^\omega$ is homeomorphic with Hilbert space.

4.210   *Proposition*
A metrizable space is second countable iff it is separable.

4.211   *Example*
The Sorgenfrey line is not metrizable.

4.212   *Theorem (Urysohn Metrization Theorem)*
Let X be a topological space.   Then the following are equivalent:
(a) X is regular, T_1, and second countable.
(b) X is normal, T_1, and second countable.
(c) X can be embedded in the Hilbert cube.
(d) X is metrizable and separable.

4.213   *Definition*
A sequence $s : \omega \rightarrow X$ in a metric space (X, d) is called **Cauchy** iff for every $\varepsilon > 0$, there exists some $N \in \omega$ such that $d(s(i), s(j)) < \varepsilon$ whenever $i, j > N$.

A metric space is said to be **complete** iff each of its Cauchy sequences converges.

4.214   *Examples*

(a) **R** with the usual metric is complete.

(b) **Q** with the usual metric is not complete.

(c) "Complete metric space" is not a topological property. There are metrics d and e on **R**, each of which induces the usual topology but such that $(\mathbf{R}, d)$ is complete and $(\mathbf{R}, e)$ is not.

4.215   *Definition*

A metrizable space (X, τ) is called **topologically complete** iff there exists *some* metric d on X such that (X, d) is complete and $\tau = \tau_d$.

4.216   *Theorem (Baire Category Theorem, II)*

Each topologically complete metrizable space is a Baire space.

4.217   *Example*

There is no continuous function $f: \mathbf{R} \to \mathbf{R}$ such that $f[\mathbf{Q}] \subseteq \mathbf{R} - \mathbf{Q}$ and $f[\mathbf{R} - \mathbf{Q}] \subseteq \mathbf{Q}$.

APPENDIX A TRANSFINITE DEFINITION OF A FUNCTION

It occasionally happens that it is necessary to *define* a function by a transfinite process. This procedure can most often be avoided; it can always be avoided in this text.

The idea, however, is the following. Suppose we wish to define a function $f: \omega \to X$ for some set X, having a particular property; we say "let $f(0) = \cdots$ and if $f(m)$ has been defined for all $m < n$, then let $f(n) = \cdots$ (usually having to do with the values $f(m)$ for $m < n$)". What we are really doing is specifying a function

$$G : \mathscr{P}(\omega \times X) \to X$$

and hoping to obtain a function $f: \omega \to X$ such that

$$f(n) = G(\{(m, f(m)) \mid m < n\}).$$

Theorem
Let $\mathscr{A}$ be a class and let $\mathscr{G} : \mathscr{P}(\mathfrak{O} \times \mathscr{A}) \to \mathscr{A}$. Then there exists a unique function $\mathscr{F} : \mathfrak{O} \to \mathscr{A}$ such that for all $B \in \mathfrak{O}$,

$$\mathscr{F}(B) = \mathscr{G}(\mathscr{F} \mid B).$$

APPENDIX B ppf CONVERSION TABLE

This table, when used in conjunction with Appendix C, can be used to reduce standard complex set-theoretic statements, step by step, to ppfs.

ppf	Notation	Verbalized
	$=$	equals
	$\in$	is an element of
	$\exists$	there exists
	$\forall$	for all
	$\ni$	such that
	$\Rightarrow$	implies
	$\mathscr{A}$	class $\mathscr{A}$
	A	set A
$\mathscr{A} \in \mathfrak{U}$		$\mathscr{A}$ is a set
not ($\mathscr{A}$ is a set)		$\mathscr{A}$ is a proper class
not ($\mathscr{A} \in \mathscr{B}$)	$\mathscr{A} \notin \mathscr{B}$	$\mathscr{A}$ is not an element of $\mathscr{B}$
$(P \Rightarrow Q)$ and $(Q \Rightarrow P)$	P iff Q	statement P holds if and only if statement Q holds
$\forall C(C \in \mathscr{A} \Rightarrow C \in \mathscr{B})$	$\mathscr{A} \subseteq \mathscr{B}$	$\mathscr{A}$ is contained in $\mathscr{B}$
not ($\mathscr{A} \subseteq \mathscr{B}$)	$\mathscr{A} \nsubseteq \mathscr{B}$	$\mathscr{A}$ is not contained in $\mathscr{B}$
$\mathscr{R} \subseteq \mathscr{A} \times \mathscr{B}$		$\mathscr{R}$ is a relation from $\mathscr{A}$ to $\mathscr{B}$
$\mathscr{R} \subseteq \mathscr{A} \times \mathscr{A}$		$\mathscr{R}$ is a relation on $\mathscr{A}$
$\exists A \ni (P \text{ and } (\exists B \ni P) \Rightarrow A = B)$	$\exists! A \ni P$	There exists a unique A such that statement P holds
$(\mathscr{F} \subseteq \mathscr{A} \times \mathscr{B})$ and $(\forall C \in \mathscr{A}, \exists! D \ni (C, D) \in \mathscr{F})$	$\mathscr{F} : \mathscr{A} \to \mathscr{B}$	$\mathscr{F}$ is a function from $\mathscr{A}$ to $\mathscr{B}$
$\mathscr{F}$ is a function and $(A, B) \in \mathscr{F}$	$B = \mathscr{F}(A)$	B is the value of $\mathscr{F}$ at A
$(\mathscr{F} : \mathscr{A} \to \mathscr{B})$ and $\mathscr{F}^{-1} \circ \mathscr{F} = 1_{\mathscr{A}}$		$\mathscr{F}$ is an injective function from $\mathscr{A}$ to $\mathscr{B}$
$(\mathscr{F} : \mathscr{A} \to \mathscr{B})$ and $\mathscr{F} \circ \mathscr{F}^{-1} = 1_{\mathscr{B}}$		$\mathscr{F}$ is a surjective function from $\mathscr{A}$ to $\mathscr{B}$

ppf	Notation	Verbalized
$\mathscr{F}$ is an injective and surjective function from $\mathscr{A}$ to $\mathscr{B}$		$\mathscr{F}$ is a bijection from $\mathscr{A}$ to $\mathscr{B}$
$\mathscr{F} : \mathscr{I} \to \mathfrak{U}$	$(\mathscr{F}(i))_{i \in \mathscr{I}}$	$(\mathscr{F}(i))_{i \in \mathscr{I}}$ is a family indexed by $\mathscr{I}$
$1_{\mathscr{A}} \subseteq \mathscr{R}$		$\mathscr{R}$ is reflexive on $\mathscr{A}$
$\mathscr{R} = \mathscr{R}^{-1}$		$\mathscr{R}$ is symmetric
$\mathscr{R} \circ \mathscr{R} \subseteq \mathscr{R}$		$\mathscr{R}$ is transitive
$\mathscr{R} \cap \mathscr{R}^{-1} \subseteq 1_{\pi_1[\mathscr{R}]}$		$\mathscr{R}$ is antisymmetric
$\mathscr{R}$ is a symmetric, transitive, reflexive relation on $\mathscr{A}$		$\mathscr{R}$ is an equivalence relation on $\mathscr{A}$
$(A \subseteq \mathscr{P}(B))$ and $((\bigcup A = B)$ and $(\varnothing \notin A)$ and $(\forall C, D \in A(C \neq D \Rightarrow C \cap D = \varnothing))$		A is a partition of B
$\mathscr{R}$ is an equivalence relation on $\mathscr{A}$ and $\mathscr{S} \subseteq \mathscr{A}$ and $\forall C \in \mathscr{A}$, $\exists! B \in \mathscr{S} \cap \mathscr{R}[\{C\}]$		$\mathscr{S}$ is a system of representatives for $\mathscr{R}$
$\mathscr{R}$ is a transitive and reflexive relation on $\mathscr{A}$		$\mathscr{R}$ is a quasi-order on $\mathscr{A}$
$\mathscr{R}$ is an antisymmetric quasi-order on $\mathscr{A}$		$\mathscr{R}$ is a partial order on $\mathscr{A}$
$\mathscr{R}$ is a quasi-order on $\mathscr{A}$ and $\mathscr{I} \subseteq \mathscr{A}$ and $\forall A, B((A \in \mathscr{I}$ and $(B, A) \in \mathscr{R}) \Rightarrow B \in \mathscr{I})$		$\mathscr{I}$ is an ideal of $\mathscr{A}$ with respect to $\mathscr{R}$
$\mathscr{R}$ is a quasi-order on $\mathscr{A}$ and $\mathscr{R}'$ is a quasi-order on $\mathscr{B}$ and $\mathscr{F}$ is a bijection from $\mathscr{A}$ to $\mathscr{B}$ and $\forall A, B((A, B) \in \mathscr{R}$ iff $(\mathscr{F}(A), \mathscr{F}(B)) \in \mathscr{R}')$		$\mathscr{F} : \mathscr{A} \to \mathscr{B}$ is an (order) isomorphism with respect to quasi-orders $\mathscr{R}$ on $\mathscr{A}$ and $\mathscr{R}'$ on $\mathscr{B}$
[a] $(\leq) \cup (\leq)^{-1} = \mathscr{A} \times \mathscr{A}$		$\leq$ is a total order on $\mathscr{A}$

[a] $\leq$ is a partial order on $\mathscr{A}$ and $\cdots$.

ppf	Notation	Verbalized
[a] $\mathscr{B} \subseteq \mathscr{A}$ and $C \in \mathscr{A}$ and $\forall D(D \in \mathscr{B} \Rightarrow D \leq C)$		C is an upper bound for $\mathscr{B}$ with respect to $\leq$
[a] $\mathscr{B} \subseteq \mathscr{A}$ and $C \in \mathscr{A}$ and $\forall D(D \in \mathscr{B} \Rightarrow D \geq C)$		C is a lower bound for $\mathscr{B}$ with respect to $\leq$
[a] $C \in \mathscr{A}$ and $\forall D((D \in \mathscr{A}$ and $C \leq D) \Rightarrow D = C)$		C is a maximal element for $\mathscr{A}$ with respect to $\leq$
[a] $C \in \mathscr{A}$ and $\forall D((D \in \mathscr{A}$ and $C \geq D) \Rightarrow D = C)$		C is a minimal element for $\mathscr{A}$ with respect to $\leq$
[a] $C \in \mathscr{A}$ and C is an upper bound for $\mathscr{A}$ with respect to $\leq$		C is the greatest element of $\mathscr{A}$ with respect to $\leq$
[a] $C \in \mathscr{A}$ and C is a lower bound for $\mathscr{A}$ with respect to $\leq$		C is the least element of $\mathscr{A}$ with respect to $\leq$
[a] $\forall C((C \neq \varnothing$ and $C \subseteq \mathscr{A}) \Rightarrow C$ has a least element with respect to $\leq)$		$\mathscr{A}$ is well-ordered by $\leq$
[a] $C, B \in \mathscr{A}$ and $B < C$ and $\forall D(B < D \Rightarrow C \leq D)$		C is the immediate successor of B with respect to $\leq$
[a] $C, B \in \mathscr{A}$ and $B > C$ and $\forall D(B > D \Rightarrow C \geq D)$		C is the immediate predecessor of B with respect to $\leq$
$(A, \leq)$ and $(A', \leq')$ are well-ordered sets and $\exists$ an order isomorphism $F : A \to A'$	$(A, \leq) \simeq (A', \leq')$	$(A, \leq)$ and $(A', \leq')$ are similar
$(\forall B, C((B \in \mathscr{A}$ and $C \in \mathscr{A}) \Rightarrow (B \in C$ or $C \in B$ or $B = C)))$ and $(\forall B(B \in \mathscr{A} \Rightarrow B \subseteq \mathscr{A}))$		$\mathscr{A}$ is an ordinal
$\mathscr{A}$ is an ordinal and $\mathscr{A}$ is a set		$\mathscr{A}$ is an ordinal number
$(\varnothing \in A)$ and $\forall a(a \in A \Rightarrow a \cup \{a\} \in A)$		A is a successor set
a is a member of the least successor set	$a \in \omega$	a is a natural number

[a] $\leq$ is a partial order on $\mathscr{A}$ and $\cdots$.

ppf	Notation	Verbalized
A is an ordinal number and $A \neq \emptyset$ and A has no immediate predecessor		A is a limit ordinal (number)
$\exists$ a bijection $F : A \to B$	$A \sim B$	A and B are equipotent
$\exists B \ni A = \text{card } B$		A is a cardinal number
card $A \in \omega$		A is finite
A is not finite		A is infinite
card $A \leq \omega$		A is countable
A is not countable		A is uncountable

This table, when used in conjunction with Appendix B, can be used to express the definitions of standard classes by means of the axiom of class formation.

Statement reducible to the ppf $P(X)$	Class determined by Axiom 2; i.e., $\{X \mid P(X)\}$		
	Symbol	Verbalized	Is a set[a]
$X \subseteq \mathscr{A}$	$\mathscr{P}(\mathscr{A})$	power class of $\mathscr{A}$	iff $\mathscr{A}$ is a set
$X = X$	$\mathfrak{U}$	the universe	no
$X \neq X$	$\varnothing$	the empty class	yes
$X \in \mathscr{A}$ or $X \in \mathscr{B}$	$\mathscr{A} \cup \mathscr{B}$	$\mathscr{A}$ union $\mathscr{B}$	iff $\mathscr{A}$ and $\mathscr{B}$ are sets
$\exists Y \curlywedge Y \in \mathscr{A}$ and $X \in Y$	$\bigcup \mathscr{A}$	union of $\mathscr{A}$	iff $\mathscr{A}$ is a set
$X \in \mathscr{A}$ and $X \in \mathscr{B}$	$\mathscr{A} \cap \mathscr{B}$	$\mathscr{A}$ intersect $\mathscr{B}$	if[a] $\mathscr{A}$ or $\mathscr{B}$ is a set
$\forall Y (Y \in \mathscr{A} \Rightarrow X \in Y)$	$\bigcap \mathscr{A}$	intersection of $\mathscr{A}$	iff $\mathscr{A} \neq \varnothing$
$X \in \mathscr{A}$ and $X \notin \mathscr{B}$	$\mathscr{A} - \mathscr{B}$	complement of $\mathscr{B}$ in $\mathscr{A}$	if[a] $\mathscr{A}$ is a set
$X = A$ or $X = B$	$\{A, B\}$	unordered pair of A and B	always
$X = A$	$\{A\}$	singleton A	always
$X = \{A\}$ or $X = \{A, B\}$	(A, B)	ordered pair with first coordinate A and second coordinate B	always
$\exists A, B \curlywedge (A \in \mathscr{A}$ and $B \in \mathscr{B}$ and $X = (A, B))$	$\mathscr{A} \times \mathscr{B}$	cartesian product of $\mathscr{A}$ and $\mathscr{B}$	iff $\mathscr{A} = \varnothing$ or $\mathscr{B} = \varnothing$ or $\mathscr{A}$ and $\mathscr{B}$ are sets
$X \in \mathscr{A} \times \{\varnothing\}$ or $X \in \mathscr{B} \times \{\{\varnothing\}\}$	$\mathscr{A} \uplus \mathscr{B}$	disjoint union of $\mathscr{A}$ and $\mathscr{B}$	iff $\mathscr{A}$ and $\mathscr{B}$ are sets
$\exists B \curlywedge (X, B) \in \mathscr{R}$	$\pi_1[\mathscr{R}]$	first projection of $\mathscr{R}$	if[a] $\mathscr{R}$ is a set
$\exists A \curlywedge (A, X) \in \mathscr{R}$	$\pi_2[\mathscr{R}]$	second projection of $\mathscr{R}$	if[a] $\mathscr{R}$ is a set
$\exists A, B \curlywedge ((A, B) \in \mathscr{R}$ and $X = (B, A))$	$\mathscr{R}^{-1}$	inverse of $\mathscr{R}$	iff $\mathscr{R}$ is a set
$\exists C \curlywedge (C \in \mathscr{C}$ and $(C, X) \in \mathscr{R})$	$\mathscr{R}[\mathscr{C}]$	image of $\mathscr{C}$ under $\mathscr{R}$	if[a] $\mathscr{R}$ is a set or ($\mathscr{R}$ is a function and $\mathscr{C}$ is a set)

[a] In this column "if" means "if but *not* only if".

Statement reducible to the ppf $P(X)$	Class determined by Axiom 2; i.e., $\{X \mid P(X)\}$		
	Symbol	Verbalized	Is a set[a]
$\exists C, D \dashv (C \in \mathscr{C}$ and $(C, D) \in \mathscr{R}$ and $X = (C, D))$	$\mathscr{R} \mid \mathscr{C}$	$\mathscr{R}$ restricted to $\mathscr{C}$	if[a] $\mathscr{R}$ is a set or ($\mathscr{R}$ is a function and $\mathscr{C}$ is a set)
$\exists A, B, C \dashv (X = (A, C)$ and $(A, B) \in \mathscr{R}$ and $(B, C) \in \mathscr{S})$	$\mathscr{S} \circ \mathscr{R}$	composition of $\mathscr{R}$ with $\mathscr{S}$	if[a] $\mathscr{R}$ and $\mathscr{S}$ are sets
$\exists A \dashv (A \in \mathscr{A}$ and $X = (A, A))$	$1_{\mathscr{A}}$ or $\Delta_{\mathscr{A}}$	identity or diagonal of $\mathscr{A}$	iff $\mathscr{A}$ is a set
$(A, X) \in \mathscr{R}$	$\mathscr{R}[\{A\}]$	equivalence class of A with respect to the equivalence relation $\mathscr{R}$	if[a] $\mathscr{R}$ is a set or $\mathscr{R}$ is a function
$\exists B \dashv (B \in A$ and $X = R[\{B\}])$	A/R	A modulo R	always
$\exists B \dashv (B \in A$ and $X = (B, R[\{B\}]))$	η_A	quotient map of A to A/R	always
$\exists A, B \dashv (X = (A, B)$ and $\mathscr{G}(A) = \mathscr{G}(B))$		kernel of the function $\mathscr{G}$	iff $\pi_1[\mathscr{G}]$ is a set
$X : \mathscr{I} \to \bigcup \{A_i \mid i \in \mathscr{I}\}$ and $\forall i (i \in \mathscr{I} \Rightarrow X(i) \in A_i)$	ΠA_i	cartesian product of $(A_i)_{i \in \mathscr{I}}$	always
$\exists F \dashv (F \in \Pi A_i$ and $X = (F, F(j)))$	π_j	jth projection function from ΠA_i	always
$\exists i, B \dashv (i \in \mathscr{I}$ and $B \in A_i$ and $X = (B, i))$	$\sum A_i$	disjoint union of $(A_i)_{i \in \mathscr{I}}$	if[a] $\mathscr{I}$ is a set
$\exists B \dashv (B \in A_j$ and $X = (B, (B, j)))$	μ_j	jth injection function into $\sum A_i$	always
$X < Y$ and $X \in \mathscr{A}$ (where $\mathscr{A}$ is well-ordered by $\leq$)	$\mathscr{S}_Y$	initial segment of $\mathscr{A}$ with respect to $\leq$ determined by Y	if[a] $\mathscr{A}$ is a set

[a] In this column, "if" means "if but *not* only if".

Statement reducible to the ppf $P(X)$	Class determined by Axiom 2; i.e., $\{X \mid P(X)\}$		
	Symbol	Verbalized	Is a set
X is an ordinal number	$\mathfrak{O}$	class of ordinal numbers	no
X is a cardinal number	$\mathfrak{C}$	class of cardinal numbers	no
$X \in \mathfrak{O}$ and $X \subseteq A$	A^+	successor of ordinal $A \in \mathfrak{O}$	always
$\forall Z((Z \in \mathfrak{O}$ and $(Z, \subseteq) \simeq (Y, \leq)) \Rightarrow X \in Z)$	$\operatorname{ord}(Y, \leq)$	ordinal number of $(Y, \leq)$ $((Y, \leq)$ a well-ordered set)	always
$\forall S(S$ is a successor set $\Rightarrow X \in S)$	ω	set of natural numbers	yes
$\forall B((B \in \mathfrak{O}$ and $B \sim A) \Rightarrow X \in B)$	$\operatorname{card} A$	cardinal number of A	always
$\exists C \dashv (C$ is a cardinal number and $\operatorname{ord}(\{A \mid A$ is an infinite cardinal number and $A < C\}) = B$ and $X \in C)$	$\aleph_B$	aleph-sub-B	always
$\forall B((B \in \mathfrak{O}$ and B uncountable) $\Rightarrow X \in B)$	Ω	first uncountable ordinal	yes

HINTS

1.41 (b) At the crucial point, use the *uniqueness* properties and 1.34.

1.45 (b) Just reverse the arrows in the proof of 1.41(b).

1.58 If $f : \mathscr{P}(A) \to A$ is one-to-one, let $B = \{a \in A \mid a \notin f^{-1}[\{a\}]\}$. Is $f(B) \in B$?

1.60 Consider $\{\{a\} \mid a \in A\} \subseteq \mathfrak{U}$.

2.23 Let $\mathscr{F}$ be the identity function on $\mathscr{Y}$ and consider $\mathscr{F}[X]$.

2.24 $X \times Y \subseteq \mathscr{P}\mathscr{P}(X \cup Y)$.

2.55 Consider separately the cases $\mathscr{I}$ is a set and $\mathscr{I}$ is a proper class.

2.68 *For* $(d) \Rightarrow (e)$. Let Q be a choice function for $\mathscr{P}(X) - \{\varnothing\}$. Let

$$T = \{(Y, R) \mid Y \in \mathscr{P}(X) - \{\varnothing\} \text{ and } R \text{ is a total order on } Y \text{ and}$$
$$\text{if } Z \text{ is a proper ideal of } (Y, R) \text{ then } Q(X - Z) \text{ is the}$$
$$\text{least element of } Y - Z\}.$$

Then (1) if $(Y, R) \in T$, then R is a well-ordering of Y; (2) $\pi_2[T]$ is totally ordered since

$$\{Z \mid \{U \in Y \mid U \mathrel{R} Z\} = \{U \in Y' \mid U \mathrel{R'} Z\}, \text{ and } R = R' \text{ on this set}\}$$

is an ideal of Y whenever (Y, R) and (Y', R') belong to T; and finally, (3) $\bigcup(\pi_2[T])$ is the desired well-ordering of X.
For $(e) \Rightarrow (a)$. Let $(X, \leq)$ be a partially ordered set with a totally ordered subset T; let $\preccurlyeq$ be a well-ordering of X in which T is an ideal. For $Y \subseteq X$ and $a \in Y - T$, let

$$B(a, Y) = \{b \in X - T \mid (c \in Y \text{ and } c \prec a) \text{ implies } b < c \text{ or } c < b\}.$$

Call a subset Y of X *admissible* iff
(1) Y is totally ordered by $\leq$,
(2) $Y \supseteq T$, and
(3) If $a \in Y - T$, then a is the $\preccurlyeq$-least element of $B(a, Y)$.
Let P be the set of admissible sets. Then P is totally ordered by $\subseteq$; $\bigcup P$ is admissible; and $\bigcup P$ is the desired maximal totally ordered set.

2.81 Use Zorn's lemma on the class of all isomorphisms from an initial segment of X to an initial segment of $\mathscr{Y}$, partially ordered by inclusion.

2.87 In the verification of (b), first show that there can be no membership loops of the form $y \in z \in x \in y$ (see 2.29).

2.88 Apply the axiom of regularity to $\mathscr{B} - \mathscr{A}$.

2.89 Consider $\mathscr{A} \cap \mathscr{B}$.

2.137 Define $f : \aleph_0 \times \aleph_0 \to \aleph_0$ by

$$f((n, m)) = \frac{(n + m)(n + m + 1)}{2} + m \quad [\textit{see } 2.129(\text{e})].$$

This formula comes from the pattern

$$(0, 0)$$
$$(1, 0) \quad (0, 1)$$
$$(2, 0) \quad (1, 1) \quad (0, 2)$$
$$(3, 0) \quad (2, 1) \quad (1, 2) \quad (0, 3)$$
$$\cdots$$

where f applied to the pattern is

$$0$$
$$1 \quad 2$$
$$3 \quad 4 \quad 5$$
$$6 \quad 7 \quad 8 \quad 9$$
$$\cdots$$

2.140   Let A be an infinite cardinal number.   Let $S = \{f \mid f: \operatorname{dom} f \to (\operatorname{dom} f)^2$ is a bijection and $\operatorname{dom} f \subseteq A\}$, where $\operatorname{dom} f$ denotes the domain of f. S, partially ordered by $\subseteq$, has a maximal element g.   If $A \sim \operatorname{dom} g$, you're essentially finished; otherwise, proceed using the lemma.

3.43   For each n, let $T_n = \{a_m \mid m \geq n\}$, and let $t_n = \sup T_n$.   Then the sequence (t_n) is monotone decreasing.

3.50   (a) Consider decimal expansions or binary (base 2) expansions.
(b) Yes.   Consider a bijection $f: \mathbf{N} \to \mathbf{Q}$.   Take inverse images of cuts.

4.16   If $\mathscr{C}$ is a countable base, consider all pairs (U, V) of elements of $\mathscr{C}$ with $U \subseteq V$.

4.58   Let Z be the union of countably many intervals of the form $[a, b)$ and countably many open points.   Let Y be the union of countably many open intervals and countably many open points.

4.77   Let $X = \{(x, m) \mid x \in \mathbf{R}, \ m \in \omega\} \subseteq \mathbf{R}^2$.   Consider the equivalence relation S on X defined by
(a) $(x, m) \ S \ (x, m)$ for all $(x, m) \in X$.
(b) $(0, m) \ S \ (0, n)$ for all $n, m \in \omega$.

4.93   (a) Show that each element of the Cantor middle third space has a unique triadic (base 3) expansion having no ones (see 3.46).

4.98   Consider $D2^{2^{\omega}}$ where D is a two-element discrete space.

4.107   If x belongs to the open set U, then there exists some $f_\alpha \in F$ such that

$$e(x) \in \pi_\alpha^{-1}[X_\alpha - \operatorname{cl} f_\alpha[X - U]].$$

4.139   Suppose that $[a, b] = S \cup T$, where S and T are disjoint, open, and non-empty.   If $a \in S$, consider the infimum of T.   Other cases are treated similarly.

4.149   If $f: \mathbf{R} \to \mathbf{R}^2$ were a homeomorphism, then the set $f[\mathbf{R} - \{0\}]$ should not be connected.

4.151   (a) Show that the theorem is true for products of finitely many spaces.
(b) Construct a dense connected subset of the product and use 4.140.

4.154   (a) Consider $[a, b]$ with open cover $\mathscr{F}$; let

$A = \{x \in [a, b] \mid$ there exist finitely many $F \in \mathscr{F}$, say $F_1, \ldots, F_n$, such that $F_1 \cup \cdots \cup F_n \supseteq [a, x)\}$.

Let $y = \sup A$. If $y = b$, okay; otherwise, obtain a contradiction.

4.158   Use 4.128 and 4.155.

4.173   To construct βX, use the embedding theorem together with the Tychonoff product theorem and theorem 4.156 to embed X in a product of closed unit intervals, and take the closure of its image as βX.

To prove that (a) implies (b), embed Y in the same way in a product of closed unit intervals.

4.177   (c) Let K and L be disjoint closed subsets of X. Construct countable open covers $\{U_n \mid n \in \omega\}$ of K and $\{V_n \mid n \in \omega\}$ of L such that for each n, cl U_n misses L and cl V_n misses K. Let $\bar{U}_n = U_n - \bigcup\{$cl $V_m \mid m \leq n\}$ and $U = \bigcup\{\bar{U}_n \mid n \in \omega\}$.

4.178   First prove the following lemma: Let S be a dense subset of $[0, 1]$ and let $\{U_s \mid s \in S\}$ be a family of open subsets of X satisfying
(a) $s < t$ implies cl $U_s \subseteq U_t$, and
(b) $X = \bigcup\{U_s \mid s \in S\}$.
Then the function $f : X \to [0, 1]$ defined by

$$f(x) = \inf \{s \in S \mid x \in U_s\}$$

is continuous.

Next inductively construct such a family of open sets corresponding to $\{n/2^m \mid m \in \omega, 0 \leq n \leq 2^m\}$ using normality.

4.182   Consider the closed disjoint subspaces $\{\Omega\} \times \omega$ and $\Omega \times \{\omega\}$.

4.183   For "not productive", let S be the Sorgenfrey line, and consider the following construction.

Let $T = \{(x, -x) \mid x \in S\}$ and let $D = \mathbf{Q} \times \mathbf{Q}$. Assuming that $S \times S$ is normal, for each $A \subseteq T$, there are open sets $U(A), V(A)$ in $S \times S$ such that $U(A) \supseteq A$, $V(A) \supseteq T - A$, and $U(A) \cap V(A) = \emptyset$. Define $f : \mathscr{P}(T) \to \mathscr{P}(D)$ by $f(A) = D \cap U(A)$. Since f is one-to-one, this is a contradiction.

4.184   For necessity of the condition: Let H be a closed subset of X and let $g : H \to [0, 1]$ be continuous. Using Urysohn's lemma and induction, define a sequence (f_i) of continuous real-valued functions on X such that
(a) $|g(x) - f_n(x)| \leq (2/3)^n$ for all $x \in H$ and
(b) $|f_n(x) - f_{n+1}(x)| \leq (2/3)^{n+1}$ for all $x \in X$.
Define f by: $f(x) = \lim f_n(x)$, the limit of the sequence $(f_n(x))$.

4.188   Consider the disjoint union of the set of all convergent sequences in the space.

4.192   Use 4.49.

4.200   If $(A_i)_{i \in \omega}$ is a sequence of nowhere-dense sets, use induction to get a decreasing sequence $(K_i)_{i \in \omega}$ of compact subsets such that $K_i \cap$ cl $A_i = \emptyset$ for each i.

4.205   (b) Suppose Ω is metrizable. For each $n \in \omega$, there is a maximal subset A_n of Ω such that every pair of distinct points of A_n is at least $1/n$ distance apart.

Since every infinite subset of Ω has an accumulation point, each A_n is finite. By maximality of each A_n, $\bigcup\{A_n \mid n \in \omega\}$ is dense. But Ω is not separable.

4.208 Let $m \leq \omega$ and let $((X_i, d_i))_{i < m}$ be a family of metric spaces. For each i, use the lemma to obtain e_i. Define $e : \Pi X_i \times \Pi X_i \to \mathbf{R}$ by

$$e(x, y) = \sum_{i < m} \frac{1}{i!} e_i(\pi_i(x), \pi_i(y)).$$

4.212 *For* (b) $\Rightarrow$ (c). For every pair (U, V) of sets from the countable base such that $\text{cl } U \subseteq V$, let $f : X \to [0, 1]$ be a continuous function such that $f[\text{cl } U] \subseteq \{0\}$ and $f[X - V] \subseteq \{1\}$. Apply the embedding theorem.

4.214 (c) Let $d(x, y) = |x - y|$ and

$$e(x, y) = d\left(\frac{x}{1 + |x|}, \frac{y}{1 + |y|}\right).$$

Appendix A Let $\mathscr{K} = \{g \mid \text{for some } B \in \mathfrak{D}, g : B \to \mathscr{A}, \text{ and for all } C \in B, g(C) = \mathscr{G}(g \mid C)\}$.

Let $\mathscr{F} = \bigcup \mathscr{K}$.

INDEX OF SYMBOLS

INDEX